T0280640

SpringerBriefs in Applied Sciences and Technology

Continuum Mechanics

More information about this series at http://www.springer.com/series/10528

Alexander Ya. Grigorenko · Wolfgang H. Müller
Yaroslav M. Grigorenko · Georgii G. Vlaikov

Recent Developments in Anisotropic Heterogeneous Shell Theory

General Theory and Applications of Classical Theory - Volume 1

Alexander Ya. Grigorenko
S.P. Timoshenko Institute of Mechanics
National Academy of Sciences of Ukraine
Kiev
Ukraine

Yaroslav M. Grigorenko
S.P. Timoshenko Institute of Mechanics
National Academy of Sciences of Ukraine
Kiev
Ukraine

Wolfgang H. Müller
Institut für Mechanik
Technische Universität Berlin
Berlin
Germany

Georgii G. Vlaikov
Technical Center
National Academy of Sciences of Ukraine
Kiev
Ukraine

ISSN 2191-530X ISSN 2191-5318 (electronic)
SpringerBriefs in Applied Sciences and Technology
ISBN 978-981-10-0352-3 ISBN 978-981-10-0353-0 (eBook)
DOI 10.1007/978-981-10-0353-0

Library of Congress Control Number: 2015958914

Printed on acid-free paper

This Springer imprint is published by SpringerNature
The registered company is Springer Science+Business Media Singapore Pte Ltd.

I have no satisfaction in formulas unless I feel their numerical magnitude.

William Thomson–Lord Kelvin

In memoriam of
Professor L. Librescu

Preface

The theory of shells is an independent and highly developed science, logically based on the theory of elasticity. Constructions consisting of thin-walled elements have found widespread applications in mechanical engineering, civil and industrial construction, ships, planes, and rockets building, as well as transport systems. The development of different shell models requires the application of hypotheses based on elasticity theory leading to a reduction in terms of two-dimensional equations that describe the deformation of the shell's middle surface. The solution of shell problems requires use of various numerical methods and involves great difficulties of computational nature. The authors present discrete–continuum approaches which they developed for solving problems of elasticity theory and which allow to reduce the initial problem to systems of ordinary differential equations. These are then solved by the stable numerical method of discrete orthogonalization and wlll be presented in this book. On the basis of these approaches, a solution for a wide class of problems of stationary deformation of anisotropic heterogeneous shells is obtained.

Many structural elements of present-day engineering, manufactured in the form of shells of various shape and complex structure with different kinds of fixation, are under the action of distributed and local loads. The wide use of shell-like elements can be attributed to the wish for satisfying the requirements associated with complex operating conditions of machines, flying vehicles, different structures, and other aggregates [7, 10–13, 15]. The complication of construction and of the structural forms of shell-like members necessitates developing a corresponding theory and methods for solving static and dynamic problems of shells made of anisotropic inhomogeneous materials.

Allowing for the aforesaid, this monograph considers approaches to solving different classes of static problems in linear formulations for anisotropic inhomogeneous shells. In this context, classic theory and different refined models are used that take specific features of shells made of modern composite materials into account. In a number of cases, the problems are solved in spatial formulation. This gives us the possibility to analyze thick-walled inhomogeneous shells and allows us to estimate the applicability of the applied theories.

In order to solve important classes of problems from the applied point of view and by taking into account the necessity of their effective realization, it seems to be expedient to use different shell models based on some simplifying assumptions. In this context, the wide use of classic theory based on the hypothesis of the invariability of the normal should be noted. It can be explained by a rather simple mathematical formulation of initial relations and by the fact that a great deal of the used shell elements possess parameters for which use of this hypothesis can be assumed to be feasible. Such circumstances make it necessary to develop methods for solving different complicated classes of problems within the framework of this hypothesis. In the case of shells made of modern composite materials, for which anisotropy and inhomogeneity in mechanical properties are typical, as well as of thick-walled shells, and shells subjected to action of local loads, it becomes necessary to take the effect of transverse strains and stresses into account, which are neglected by the classic theory. The efficiency of the following realization should be taken into account when constructing a refined shell theory. It is directly connected with a mechanical interpretation of the adopted assumptions, with the simplicity of mathematical formulation of initial relations, and with the order of the resolving equations.

A special feature during the development of plate and shell theories consists of connecting a mathematical model for a specific class of problems and developing the corresponding solution method. This fact is supported by the Kirchhoff–Love theory of thin shells [8, 9], where the objective is pursued by describing adequately the behavior of plates and shells and by retaining the simplicity of these models to such an extent that it is possible to solve problems with existing computing facilities. This interconnection becomes all the more evident in present times, when computers are widely used for solving shell theory problems.

> The necessity for obtaining numerical answers to numerically posed questions is a central stimulus for origin of new theories during all the path mathematics was developed. One may say with confidence that the most part of the science, which is named today as classical applied mathematics, is originated due to the desire to get rid of laborious calculations. Now, when the possibilities of computers are increased significantly, the center of gravity has shifted: the emphasis is on the search of the most efficient and convenient methods for their performing instead of total relieving of having to do arithmetic calculations (Casti and Kalaba [2]).

> In order to ensure that the powerful computational potentialities, which we have at our disposal, would be exploited, it is necessary to carry out bulky preliminary analysis of problem statements and analytical methods making some sense in application. If previously researchers were striving to simplify the problems so the linear functional equations would be obtained, at the present time an objective is to reduce computation problems to the Cauchy problems for ordinary differential equations (linear or nonlinear). As soon as a physical, economical, technical, or biological problem is reduced to the Cauchy problem for ordinary differential equations, its total solution may be thought of as approaching to completion (Bellman and Calaba [6]).

Along with universal approaches to solving problems of mechanics and mathematical physics, based on using finite elements [3, 4, 14, 16, 17], boundary element [1, 5], and other discrete methods, nowadays approaches that allow to reduce a problem to ordinary differential equations, based on approximation by other variables, using analytical tools, have found their wide application to solving certain classes of problems.

The approaches proposed are realized in computational complexes using stable numerical methods. It allows us solving problems with high accuracy for shells of different shape and structure in varying parameters in a wide range. Possibilities of the approaches proposed are supported by the examples of the number of complicated problems including analysis of specific structural elements which are widely used in different engineering applications.

In this context, the present monograph represents some approaches to numerical–analytical methods for solving the problems of mechanics of shells with various structure and shape based on the classical, refined, and spatial models.

The monograph consists of two books, each of which consists of three chapters. A summary of the chapters is as follows.

Book 1

Chapter 1: Elastic bodies in the form of thin- and thick-walled anisotropic shells are considered. The shells may be made of both homogeneous and inhomogeneous materials with discrete (multilayer) structure or of continuously inhomogeneous materials (functionally gradient materials). The stationary deformation of shells of the above class is analyzed by using various mechanical models. The basic relations of the theory of elasticity, which include the equations of equilibrium, geometrical, and physical relations, are presented. By using classical and refined shell theories, the original three-dimensional problem is reduced to a two-dimensional one. The fundamental equations of the classical (Kirchhoff–Love) shell theory, which are based on the hypothesis of undeformed normals, are presented. It is assumed that all of the shell layers are stiffly joined and operate mutually without sliding and separation. It is assumed that geometrical and mechanical parameters of the shells and mechanical loads applied to them are such that when considering the shell as a unit stack, the hypothesis of undeformed normals is valid. In case of laminated shells made of new composite materials with low shearing stiffness, essential anisotropy, and inhomogeneity of mechanical properties, whose characteristics of layers are highly dissimilar, the refined model based on the straight-line hypothesis is used. The basic equations of the model are presented. The various boundary physically consistent conditions at the bounding surfaces of the shells are specified.

Chapter 2: The numerical–analytical methods for solving boundary-value and boundary-value eigenvalue problems for the systems of ordinary differential equations and partial differential equations with variable coefficients are presented.

In order to solve one-dimensional problems, the discrete-orthogonalization method is proposed. Such an approach is based on reducing the boundary-value problem to a number of Cauchy problems and on their orthogonalization at some points of the integration interval which provides stability of calculations. In case of boundary-value eigenvalue problems, such an approach is employed in combination with an incremental search method. In order to solve two-dimensional problems, an approach based on reducing the original partial system to systems of ordinary differential equations by making use of spline approximation, solved by the discrete-orthogonalization method, is proposed. Employing spline-functions has the following advantages: Stability with respect to local disturbances, i.e., spline behavior in the vicinity of a point does not influence the spline behavior as a whole, as it does, for example, in polynomial approximation; more satisfactory convergence in contrast to the case of polynomials being applied as approximation functions; simplicity and convenience in calculation and implementation of spline-functions with the help of modern computers. Besides that a nontraditional approach to solving problems of the above class is proposed. The approach employs discrete Fourier series, i.e., Fourier series for functions specified on the discrete set of points. The two-dimensional boundary-value problem is solved by reducing it to a one-dimensional one as a result of introducing auxiliary functions and by separation of variables when using discrete Fourier series. Taking into account the calculation possibilities of modern computers, which make it possible to calculate a large number of series terms, the problem can be solved with high accuracy.

Chapter 3: The results of studying stationary deformation of anisotropic inhomogeneous shells of various classes by using the classical Kirchhoff–Love theory and numerical approaches outlined in the Chap. 2 of the present book are presented. The stress–strain problems for shallow, noncircular cylindrical shells and shells of revolution are solved. The various types of boundary conditions and loadings are considered. Distributions of stress and displacement fields in the shells of the above classes are analyzed for various geometrical and mechanical parameters. The practically important stress problem for a high-pressure glass-reinforced balloon is solved. The dynamical characteristics of an inhomogeneous orthotropic plate under various boundary conditions are studied. The problem of free vibrations of a circumferential inhomogeneous truncated conical shell is solved. The effect of variation in thickness, mechanical parameters, and boundary conditions on the behavior of natural frequencies and vibration modes of a plate and cone is analyzed. Much attention is given to validation of the reliability of the results obtained by numerical calculations.

Book 2

Chapter 1: The solutions of stress–strain problems for a wide class of anisotropic inhomogeneous shells obtained by the refined model are presented. Studying these problems results in the calculations of severe difficulties due to partial differential equations with variable coefficients. For solving the problem, spline-collocation and discrete-orthogonalization methods are used. The influence of geometrical and mechanical parameters, of the boundary conditions, and of the loading character on the distributions of stress and displacement fields in shallow, spherical, conical, and noncircular cylindrical shells is analyzed. The dependence of the stress–strain pattern on shell thickness variation is studied. The problem was solved also in the case of the thickness varying in two directions. It is studied how the rule of variation in the thickness of the shells influences their stress–strain state. Noncircular cylindrical shells with elliptical and corrugated sections are considered.

The results obtained in numerous calculations support the efficiency of the discrete-orthogonalization approach proposed in the monograph for solving static problems for anisotropic inhomogeneous shells when using the refined model.

Chapter 2: A wide class of problems of natural vibrations of anisotropic inhomogeneous shells is solved by using a refined model. Shells with constructional (variable thickness) and structural inhomogeneity (made of functionally gradient materials) are considered. The initial boundary-value eigenvalue partial derivative problems with variable coefficients are solved by spline-collocation, discrete-orthogonalization, and incremental search methods. In case of hinged shells, the results obtained by making use of analytical and proposed numerical methods are compared and analyzed. It is studied how the geometrical and mechanical parameters as well as the type of boundary conditions influence the distribution of dynamical characteristics of the shells under consideration. The frequencies and modes of natural vibrations of an orthotropic shallow shell of double curvature with variable thickness and various values of curvature radius are determined. For the example of cylindrical shells made of a functionally gradient material, the dynamical characteristics have been calculated with the thickness being differently varied in circumferential direction. The values of natural frequencies obtained for this class of shells under some boundary conditions are compared with the data calculated by the three-dimensional theory of elasticity.

Chapter 3: The model of the three-dimensional theory of elasticity is employed in order to study stationary deformation of hollow anisotropic inhomogeneous cylinders of finite length. Solutions of problems of the stress–strain state and natural vibrations of hollow inhomogeneous finite-length cylinders are presented, which were obtained by making use of spline-collocation and discrete-orthogonalization methods. The influence of geometrical and mechanical parameters, of boundary conditions, and of the loading character on distributions of stress and displacement fields, as well as of dynamical characteristics in the above cylinders is analyzed. For some cases, the results obtained by three-dimensional and shell theories are compared. When solving dynamical problems for orthotropic hollow cylinders with different boundary conditions at the ends, the method of straight-line methods in combination with the discrete-orthogonalization method was also applied. Computations for solid anisotropic finite-length cylinders with different end conditions were carried out by using the semi-analytical finite element method. In case of free ends, the results of calculations the natural frequencies were compared with those determined experimentally. The results of calculations of mechanical behavior of anisotropic inhomogeneous circular cylinders demonstrate the efficiency of the discrete–continual approaches proposed in the monograph for solving shell problems using the three-dimensional model of the theory of elasticity.

References

1. Banerjee PK (1994) The boundary element methods in engineering. Mc Graw-Hill College
2. Bellman R, Kalaba R (1965) Quasilinearization and nonlinear boundary-value problems. Elsevier, Amsterdam
3. Bernadou M (1996) Finite element methods for thin shell problems. Wiley, New York
4. Bischoff M, Wall WA, Bletzinger KU, Ramm E (2004) Models and finite elements for thin-walled structures. In: Stein E, De Borst R, Hughes TJR (eds) Encyclopedia of computational mechanics. Volume 2: solids and structures. Wiley, Chichester, pp 59–137
5. Brebbia CA, Walker S (1984) Boundary element technique in engineering. Buttterworth
6. Casti J, Kalaba R (1973). Imbedding methods in applied mathematics. Addison-Wesley, Reading, MA
7. Farshad M (1992) Design and analysis of shell structures farshad of Plate and Shell Structures. Springer, Berlin
8. Love AEH (1952) Mathematical theory of elasticity. Cambridge University Press, Cambridge
9. Love AEH (1888) On the small free vibrations and deformations of thin elastic shells. Philos Trans R Soc Lond Ser A 179:491–546
10. Maan HJ (2003) Design of plate and shell structures. Wiley, New York
11. Mungan I, Abelj F (2011) Fifty years of progress for shell and spatial structures. Multi Science Publishing Co Ltd
12. Pietraszkiewicz W (2014) Shell-structures: theory and application. CRC Press, Boca Raton
13. Ramm E, Wall WA (2005) Computational Methods for Shells. Special Issue of Comput Methods Appl Mech Eng 194:2285–2707
14. Reddy JN (2005) An introduction to the finite element method. McGraw-Hill Education, New York
15. Reddy JN (2007) Theory and analysis of elastic plates and shells. CRC Press, Taylor and Francis, Boca Raton
16. Zienkiewicz OC, Taylor RL (1989) .The finite element method. McGraw-Hill, New York
17. Zienkiewicz OC, Taylor RL, Too JM (1971). Reduced integration technique in general analysis of plates and shells. Int J Numer Methods Eng 3:275–290

Contents

Chapter 1
Mechanics of Anisotropic Heterogeneous Shells: Fundamental Relations for Different Models

Abstract Elastic bodies in the form of thin and thick-walled anisotropic shells are considered. The shells may be made both of homogeneous and inhomogeneous materials with discrete (multilayer) structure or of continuously inhomogeneous materials (functionally gradient materials). The stationary deformation of such shells is analyzed by using various mechanical models. The basic relations of the theory of elasticity, which include equilibrium equations of motion, geometrical, and physical relations, are presented. By using classical and refined shell theories, the original three-dimensional problem is reduced to a two-dimensional one. The basic equations of the classical (Kirchhoff-Love) shell theory, which are based on the hypothesis of undeformed normals, are presented. It is assumed that all of the shell layers are stiffly joined and operate mutually without sliding and separation. Moreover, geometrical and mechanical parameters of the shells and mechanical loads applied to them are such that, considering the shell as a unit stack, the hypothesis of undeformed normals is valid. In the case of laminated shells made of new composite materials with low shearing stiffness, where anisotropy and inhomogeneity of the mechanical properties of the layers vary considerably, the refined model based on the straight-line hypothesis is used. The basic equations of the model are presented and various physically consistent boundary conditions at the bonded surfaces of the shells are specified.

Keywords 3D elasticity theory · Classical shell models · Refined shell models · Basic relations

1.1 Introduction

The present chapter presents the fundamental equations of linear elasticity theory as pertinent to classical and refined shell theory. Shell theory, which attracts much attention due to its wide application for solving practical engineering problems, can be considered as a part of (linear) elasticity theory, which is described, for example

A.Ya. Grigorenko et al., *Recent Developments in Anisotropic Heterogeneous Shell Theory*, SpringerBriefs in Continuum Mechanics,
DOI 10.1007/978-981-10-0353-0_1

in the books by Love [17], Timoshenko and Goodier [26], Sokolnikoff [25], or Nowacki [21]. A general, not to mention a closed-form solution of shell problems is impossible within the framework of elasticity theory. However, in some cases, in particular for shells of cylindrical shape, this theory makes it possible to examine stationary deformations. *Therefore, in what follows, we will concentrate on the fundamental time-independent relations.* The first simple models of thin shell theory appeared in Love's and Lamb's work on elasticity theory [13, 17]. However, in the case of thick-walled shells made of modern composite materials, which reveal anisotropy and inhomogeneity in their physical and mechanical properties, transverse stresses and strains must be taken into account. These were neglected in the classical theory. The work of Timoshenko and Woinowsky-Krieger [27], Mindlin [18], and Reissner et al. [23] can be considered as major breakthrough in this area of research. The theory of shells has a long history caused by the demands for creating structures and constructions of different design. It is a special branch of the mechanics of a deformable body and still under rapid development. The various, quite extensive results obtained in making, developing, and applying the theory of shells are of fundamental as well as of applied character.

Fundamental theoretical results in shell theory have been obtained by various scientists of the highest reputation in the mechanics community, such as Timoshenko and Woinowsky-Krieger [27], Flügge [7, 8], Donell [6], Leissa [14], Kraus [12], Gol'denveizer [9], Mushtari and Galimov [19], Novozhilov [20], Vlasov [28], etc.

Due to the wide application of new composite materials, the theory of laminated anisotropic shells came to the attention of many researchers. To a large part its development must be attributed to Altenbach and Eremeyev [1], Ambartsumyan [2–4], Burton and Noor [5], Librescu [15], Reddy [22], Lechnitskii [16].

We will consider stationary deformation of thick-walled as well as thin-walled shell constructions. Depending on the geometrical and mechanical parameters of a shell structure, we will concentrate our attention initially on the classical shell theory and refined spatial models.

Moreover, thin flexible shells will be considered. The shells may be both single-layered and multilayered, i.e., assembled of an arbitrary number of inhomogeneous anisotropic layers whose rigidity is variable in two coordinate directions. We suppose that all layers of the package deform without slipping and separation. The material of the shell obeys Hooke's law. It is assumed that geometrical and mechanical parameters of the shell, the ways of its fixation and the loading are such that in order to describe the deformation process, geometrically nonlinear theory of thin shells in quadratic approximation may be used (Grigorenko and Gulyaev [10], Sanders [24]). This theory is based on the hypothesis of undeformed normals for the whole package of layers. The essence of the Kirchhoff-Love hypothesis is that the elongation in the direction of the normal to a coordinate surface and the transverse shears are assumed to be approximately zero. Moreover, the normal stresses on the areas parallel to the coordinate surface are neglected in comparison with the analogous stresses on the areas perpendicular to the coordinate surface. The use of the hypothesis of undeformed normals makes it possible to reduce the problem of deformation of layered anisotropic shells to the problem of deformation of a coordinate surface and to solve a two-dimensional problem instead of a three-dimensional one.

In the case of homogeneous isotropic shells, a midsurface is typically chosen as coordinate one. For sandwich anisotropic shells this surface is chosen in such a way that the notation of the elasticity relations becomes as simple as possible.

1.2 Initial Assumptions

1.2.1 Curvilinear Orthogonal Coordinate System

Besides Cartesian coordinate systems, curvilinear ones are frequently used in elasticity theory. The use of curvilinear coordinates is useful both in view of the shape of the bounding surfaces for the considered classes of elastic bodies and the potential for solving the associated boundary-value problems of mathematical physics.

The position of an arbitrary point M of space (see Fig. 1.1) in a curvilinear coordinate system is unambiguously determined by three numbers, α, β, γ. The values referring to curvilinear coordinates are related to the Cartesian coordinates x, y, z by bijective functions:

$$\alpha = \alpha(x, y, z), \quad \beta = \beta(x, y, z), \quad \gamma = \gamma(x, y, z), \tag{1.2.1}$$

i.e., because the position of the point P in space is well defined, if the values α, β, γ are specified, inverse dependencies must exist:

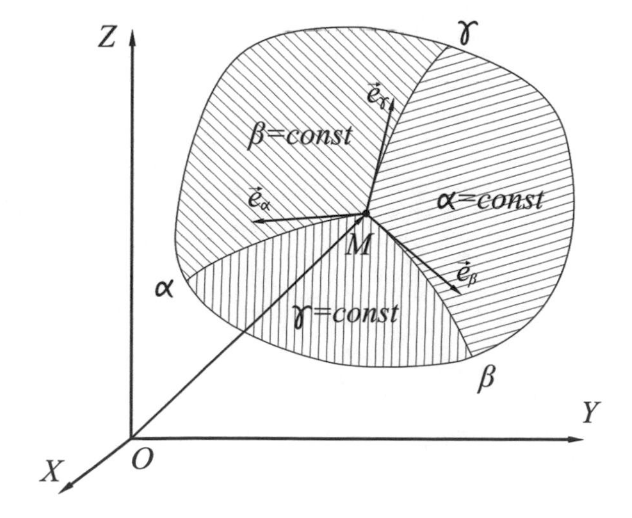

Fig. 1.1 Labeling points in space

$$x = x(\alpha, \beta, \gamma), \quad y = y(\alpha, \beta, \gamma), \quad z = z(\alpha, \beta, \gamma), \tag{1.2.2}$$

where x, y, z are unique defined in terms of the parameters α, β, γ. They are C^1 continuous, i.e., with a continuous first derivative.

The equations $\alpha = \text{const.}$, $\beta = \text{const.}$, $\gamma = \text{const.}$. define three families of surfaces, which will be referred to as coordinate surfaces. Only one surface of each family runs through each point M of space. Moreover, three coordinate lines, which are intersections of associated coordinate surfaces, run through the point M. The coordinate line α is formed by intersection of the coordinate surfaces $\beta = \text{const.}$ and $\gamma = \text{const.}$, the coordinate line β by intersection of coordinate surfaces $\alpha = \text{const.}$ and $\gamma = \text{const.}$, and the line γ by intersection of the surfaces $\alpha = \text{const.}$ and $\beta = \text{const.}$, respectively.

In many cases it is more convenient to describe a surface by using parametric equations. For this purpose, the concept of surface coordinates is introduced.

Let the one-parameter set of lines be such that each line is characterized by a specific value of a certain parameter given at some surface. A reference grid is considered to be prescribed on the surface if two such sets are specified on the surface in such a way that each line from the first set intersects with each line of the second set at the common point. Let the lines of the first set generating the coordinate grid be defined by values of some parameter ξ, while the lines of the second set are defined by values of the parameter η (Fig. 1.2). Since each point of the surface is intersected by only one curve from each set, the location of the points on the surface is identifiable by the parameters ξ and η. These parameters are called coordinates on the given surface. The surface with embedded coordinates ξ and η is referred to as a parametrized surface.

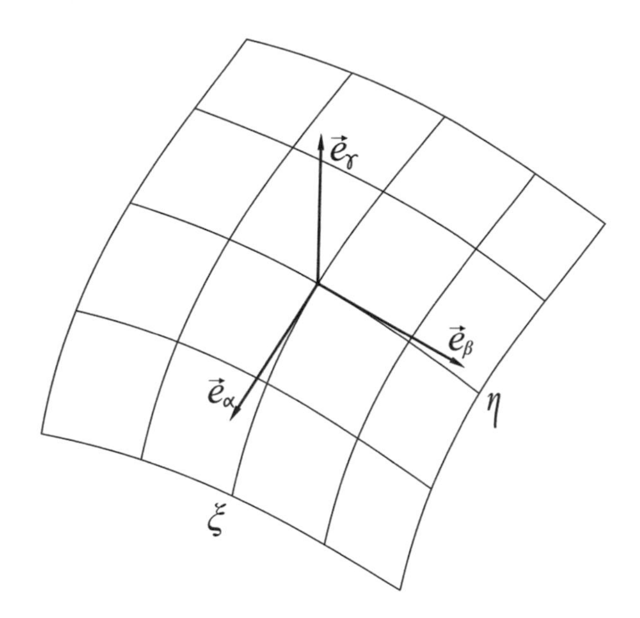

Fig. 1.2 Surface coordinates (see text)

It is clear that any point of a surface can also be characterized by Cartesian coordinates x, y, z. Therefore, the Cartesian coordinates of the points on the parametrized surface are functions of the coordinates on the surface:

$$x = x(\xi, \eta); \quad y = y(\xi, \eta); \quad z = z(\xi, \eta), \tag{1.2.3}$$

where the functions x, y, z are single-valued and continuous with respect to the parameters ξ, η. It is assumed that the functions (1.2.3) have continuous partial derivatives for ξ and η of any desired order.

In many cases it is more advantageous to make use of the orthogonal system of curvilinear coordinates ξ, η, i.e., a system of ξ and η curves intersecting at each point at a right angle.

Let some curve, with the exception of the lines ξ and η, be prescribed on a given surface. If some parameter t is introduced on this curve each of its values will correspond to some point of the surface with coordinates ξ and η. Thus, the coordinates $\vec{e}_n\xi$ and η along the curve are functions of the parameter t, i.e., $\xi = \xi(t), \eta = \eta(t)$. These equations are the equations of a curve on a surface. By substituting them into Eq. (1.2.3), we obtain a parametric representation of an arbitrary curve on a surface. Thus, by setting successively $\xi =$ const. and $\eta =$ const. in (1.2.3), we obtain two populations of coordinate lines representing the η- and ξ-curves, respectively.

Evidently, a coordinate grid can be chosen not only in one way. Let a point M and a normal $\vec{e}_n$ be given on a surface. A bundle of planes can pass through the normal $\vec{e}_n$. Then each plane will intersect the surface in some curve. The section produced with planes passing through a normal to the surface is said to be normal. The curvature of this section can be calculated from differential geometry, which allows to show (Korn and Korn [11]), that two mutually perpendicular directions with extremal values of curvatures exist at each point of a surface. These directions are called principal directions of a surface. The curvatures of normal sections whose tangents coincide with the principal directions are called principal curvatures. If the coordinate lines coincide with the lines of principal directions, these curvatures are said to be lines of curvature.

The relations of the linear theory of elasticity become most simple in curvilinear orthogonal coordinates, i.e., in such coordinates, where the coordinate lines α, β, γ at each point M are mutually perpendicular. In such systems the expression for the length of the square of a line element has the form:

$$ds^2 = H_1^2 d\alpha^2 + H_2^2 d\beta^2 + H_3^2 d\gamma^2. \tag{1.2.4}$$

In general, the values H_1, H_2 and H_3 are functions of the coordinates α, β, γ. They are known as Lamé coefficients or Lamé parameters (not to be confused with the isotropic elastic material parameters, λ and μ, which are also named after Lamé). The Lamé parameters can be determined from:

$$H_1^2 = \left(\frac{\partial x}{\partial \alpha}\right)^2 + \left(\frac{\partial y}{\partial \alpha}\right)^2 + \left(\frac{\partial z}{\partial \alpha}\right)^2, \quad H_2^2 = \left(\frac{\partial x}{\partial \beta}\right)^2 + \left(\frac{\partial y}{\partial \beta}\right)^2 + \left(\frac{\partial z}{\partial \beta}\right)^2,$$
$$H_3^2 = \left(\frac{\partial x}{\partial \gamma}\right)^2 + \left(\frac{\partial y}{\partial \gamma}\right)^2 + \left(\frac{\partial z}{\partial \gamma}\right)^2. \tag{1.2.5}$$

Let us consider some coordinate systems, which are frequently used for solving problems of elasticity theory.

For obvious reasons cylindrical shells will be considered in a cylindrical coordinate system $\alpha = z,\ \beta = \theta,\ \gamma = r$. The coordinate surfaces consist of circular cylinders r = const., planes θ = const., running through the axis Oz, and planes z = const., perpendicular to the Oz-axis. In this case the relations (1.2.3) become:

$$x = r\cos\theta, \quad y = r\sin\theta, \quad z = z. \tag{1.2.6}$$

The Lamé coefficients are given by:

$$H_1 = 1, \quad H_2 = r, \quad H_3 = 1. \tag{1.2.7}$$

Moreover, it is convenient to consider spherical shells in a spherical coordinate system $\alpha = \theta,\ \beta = \phi,\ \gamma = r$. Now the coordinate surfaces consist of spheres $r =$ const., cones $\phi =$ const. with the apex being at the point O, and planes $\theta =$ const. running through the Oz-axis. In this case, the relations (1.2.2) will take the form:

$$x = r\sin\phi\cos\theta, \quad y = r\sin\phi\cos\theta, \quad z = r\cos\theta \tag{1.2.8}$$

and the Lamé parameters are:

$$H_1 = 1, \quad H_2 = r, \quad H_3 = r\sin\phi \tag{1.2.9}$$

respectively.

Finally note that that the Lamé parameters in a Cartesian coordinate system are simply given by $H_1 = H_2 = H_3 = 1$.

From differential geometry (Korn and Korn [11]) it is known that the Lamé coefficients are no independent functions. They have to satisfy six differential relations:

$$
\begin{aligned}
&\frac{\partial}{\partial \alpha}\left(\frac{1}{H_1}\frac{\partial H_2}{\partial \alpha}\right)+\frac{\partial}{\partial \beta}\left(\frac{1}{H_2}\frac{\partial H_1}{\partial \beta}\right)+\frac{1}{H_3^2}\frac{\partial H_1}{\partial \gamma}\frac{\partial H_2}{\partial \gamma}=0,\\
&\frac{\partial}{\partial \beta}\left(\frac{1}{H_2}\frac{\partial H_3}{\partial \beta}\right)+\frac{\partial}{\partial \gamma}\left(\frac{1}{H_3}\frac{\partial H_2}{\partial \gamma}\right)+\frac{1}{H_1^2}\frac{\partial H_2}{\partial \alpha}\frac{\partial H_3}{\partial \alpha}=0,\\
&\frac{\partial}{\partial \gamma}\left(\frac{1}{H_1}\frac{\partial H_1}{\partial \gamma}\right)+\frac{\partial}{\partial \alpha}\left(\frac{1}{H_1}\frac{\partial H_3}{\partial \alpha}\right)+\frac{1}{H_2^2}\frac{\partial H_3}{\partial \beta}\frac{\partial H_1}{\partial \beta}=0,\\
&\frac{\partial^2 H_1}{\partial \beta \partial \gamma}-\frac{1}{H_2}\frac{\partial H_2}{\partial \gamma}\frac{\partial H_1}{\partial \beta}-\frac{1}{H_3}\frac{\partial H_3}{\partial \beta}\frac{\partial H_1}{\partial \gamma}=0,\\
&\frac{\partial^2 H_2}{\partial \alpha \partial \gamma}-\frac{1}{H_3}\frac{\partial H_3}{\partial \alpha}\frac{\partial H_2}{\partial \gamma}-\frac{1}{H_1}\frac{\partial H_1}{\partial \gamma}\frac{\partial H_2}{\partial \alpha}=0,\\
&\frac{\partial^2 H_3}{\partial \alpha \partial \beta}-\frac{1}{H_1}\frac{\partial H_1}{\partial \beta}\frac{\partial H_3}{\partial \alpha}-\frac{1}{H_2}\frac{\partial H_2}{\partial \alpha}\frac{\partial H_3}{\partial \beta}=0.
\end{aligned}
\tag{1.2.10}
$$

1.2.2 Shell Geometry

We will consider shells in an undeformed state as three-dimensional bodies and choose some initial coordinate surface (plane of reference). We characterize the shell surface by an orthogonal curvilinear conjugated system of coordinates α, β, where the lines $\alpha = \text{const.}$ and $\beta = \text{const.}$ align with the lines of principal curvatures (Fig. 1.1). Suppose that the relation between the Cartesian and the curvilinear coordinates on the reference plane is given by unique relations, so that location of an arbitrary point M on the surface,

$$x = x(\alpha, \beta), \quad y = y(\alpha, \beta), \quad z = z(\alpha, \beta) \tag{1.2.11}$$

can be determined by the position vector

$$\vec{r} = \vec{r}(\alpha, \beta) = \{x, y, z\}. \tag{1.2.12}$$

Then the first quadratic form of the initial coordinate surface is (Korn and Korn [11]):

$$\phi_1 = \mathrm{d}\vec{r}^2 = \mathrm{d}s^2 = A^2\mathrm{d}\alpha^2 + B^2\mathrm{d}\beta^2. \tag{1.2.13}$$

Here A and B are the Lamé parameters. They are related to the increments of arcs of coordinate lines by:

$$\mathrm{d}s_\alpha = A\mathrm{d}\alpha, \quad \mathrm{d}s_\beta = B\mathrm{d}\beta. \tag{1.2.14}$$

They may be expressed in terms of coordinates of the position vector $\vec{r}$ as follows:

$$A^2 = \left|\frac{\partial \vec{r}}{\partial \alpha}\right|^2 = \left(\frac{\partial x}{\partial \alpha}\right)^2 + \left(\frac{\partial y}{\partial \alpha}\right)^2 + \left(\frac{\partial z}{\partial \alpha}\right)^2,$$
$$B^2 = \left|\frac{\partial \vec{r}}{\partial \beta}\right|^2 = \left(\frac{\partial x}{\partial \beta}\right)^2 + \left(\frac{\partial y}{\partial \beta}\right)^2 + \left(\frac{\partial z}{\partial \beta}\right)^2. \quad (1.2.15)$$

The first quadratic form defines the inner geometry (Korn and Korn [11]) of the coordinate surface. This form can be used to determine the length of curves on the surface, the angles between the curves, and the surface areas.

The second quadratic form of the coordinate surface is:

$$\phi_2 = L\mathrm{d}\alpha^2 + N\mathrm{d}\beta^2. \quad (1.2.16)$$

Let us denote by |.| (Korn and Korn [11]) the scalar triple product (a.k.a. mixed product) of vectors. The coefficients L and N are defined by:

$$L = \frac{\left[\frac{\partial^2 \vec{r}}{\partial \alpha^2}\,\frac{\partial \vec{r}}{\partial \alpha}\,\frac{\partial \vec{r}}{\partial \beta}\right]}{AB}, \quad N = \frac{\left[\frac{\partial^2 \vec{r}}{\partial \beta^2}\,\frac{\partial \vec{r}}{\partial \alpha}\,\frac{\partial \vec{r}}{\partial \beta}\right]}{AB} \quad (1.2.17)$$

and related to the principal curvatures $\frac{1}{R_\alpha}$ and $\frac{1}{R_\alpha}$ by the expressions:

$$\frac{1}{R_\varepsilon} = -\frac{L}{A^2} = -\frac{1}{A^3B}\begin{vmatrix} x''_{\alpha\alpha} & y''_{\alpha\alpha} & z''_{\alpha\alpha} \\ x'_\alpha & y'_\alpha & z'_\alpha \\ x'_\beta & y'_\beta & z'_\beta \end{vmatrix},$$
$$\frac{1}{R_\beta} = -\frac{N}{B^2} = -\frac{1}{AB^3}\begin{vmatrix} x''_{\beta\beta} & y''_{\beta\beta} & z''_{\beta\beta} \\ x'_\alpha & y'_\alpha & z'_\alpha \\ x'_\beta & y'_\beta & z'_\beta \end{vmatrix}. \quad (1.2.18)$$

The curvature radii R_α and R_β can be both positive and negative depending on how the direction of the normal to the coordinate surface is chosen. The second quadratic form defines the outer geometry (Korn and Korn [11]) of the surface and characterizes the curvature of a line on the coordinate surface.

The coefficients of the first and second quadratic forms are related to each other by the following Gauss-Codazzi conditions:

$$\frac{\partial}{\partial \alpha}\left(\frac{B}{R_\beta}\right) = \frac{1}{R_\alpha}\frac{\partial B}{\partial \alpha}, \quad \frac{\partial}{\partial \beta}\left(\frac{A}{R_\alpha}\right) = \frac{1}{R_\beta}\frac{\partial A}{\partial \beta},$$
$$\frac{\partial}{\partial \alpha}\left(\frac{1}{A}\frac{\partial B}{\partial \alpha}\right) + \frac{\partial}{\partial \beta}\left(\frac{1}{B}\frac{\partial A}{\partial \beta}\right) = -\frac{AB}{R_\alpha R_\beta}. \quad (1.2.19)$$

In the general case, we will consider shells assembled of layers with variable thickness (Fig. 1.3).

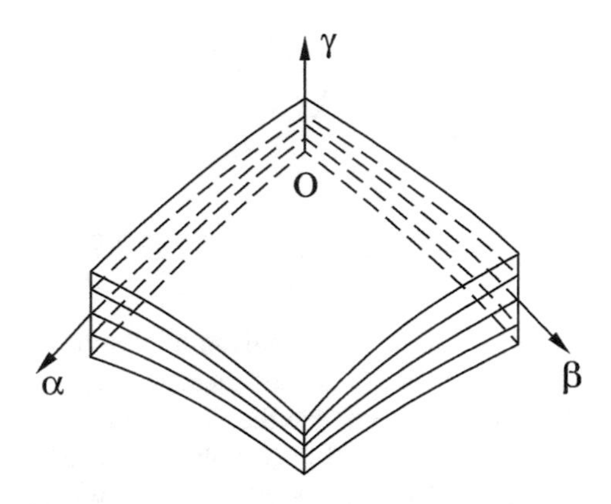

Fig. 1.3 Shells consisting of layers with variable thickness

The normal section of the shell is shown in Fig. 1.4. We will orient the coordinate γ along the normal to the coordinate surface. The thickness of the shell, h, is counted from the coordinate surface $\gamma = 0$. It is a function of α and β, i.e., $h = h\ (\alpha,\ \beta)$. In this case, the contact surfaces of the ith and $(i + 1)$th layers as well as the bounding surfaces $\gamma = \gamma_0\ (\alpha,\ \beta)$ and $\gamma = \gamma_n\ (\alpha,\ \beta)$ are specified by equations $\gamma = \gamma_i\ (\alpha,\ \beta)$ $(i \equiv 0,\ 1,\ 2,\ \ldots,\ n)$. The end surfaces of the shell coincide with the coordinate lines.

They are defined by the equations $\alpha = \text{const.}$ and $\beta = \text{const.}$ Thus, the shell being a three-dimensional body, the undeformed state is referred to the orthogonal coordinate system α, β, γ. In the chosen coordinate system we find for the differential of the arc length:

$$\mathrm{d}s^2 = H_1^2\mathrm{d}\alpha^2 + H_2^2\mathrm{d}\beta^2 + H_3^2\mathrm{d}\gamma^2, \qquad (1.2.20)$$

where $H_1,\ H_2,\ H_3$ are the corresponding Lamé coefficients:

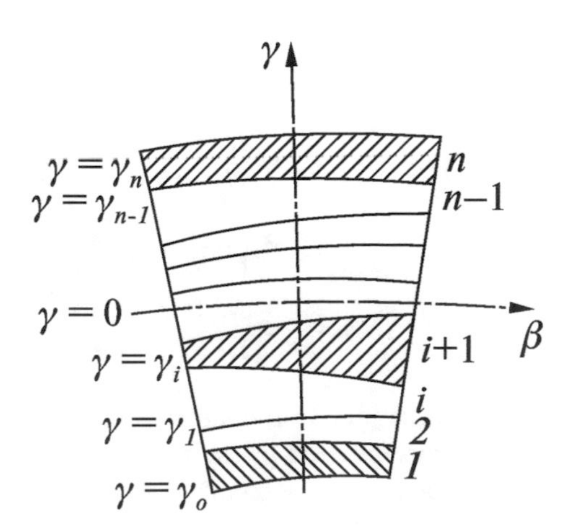

Fig. 1.4 Normal section of shell

$$H_1 = A\left(1+\frac{\gamma}{R_\alpha}\right), \quad H_2 = B\left(1+\frac{\gamma}{R_\beta}\right), \quad H_3 = 1. \tag{1.2.21}$$

1.3 Basic Relations of 3D Elasticity Theory

Let us consider elastic anisotropic bodies both continuously inhomogeneous along the normal coordinate and piecewise-continuous in the form of a stack of layers. For their study we will use the three-dimensional theory of elasticity (Lekhnitskii [16], Love [17], Novacki [21], Sokolnikoff [25], Timoshenko and Goodier [26]), which we also refer to as *spatial* theory. We assume that the displacements are small compared to the dimensions of the body and that the angles of rotation are small when compared to one. In other words, we use the equations of the linear spatial theory of elasticity. The materials being considered obey the generalized Hooke's law, i.e., stress and strain components are related linearly. In the general case, the coefficients of these dependencies are variable functions of the coordinates.

Let us denote the components of the displacement vector $\vec{u}(\alpha,\beta,\gamma,t)$ in the direction of the basis $\vec{e}_\alpha$, $\vec{e}_\beta$, $\vec{e}_\gamma$ of the curvilinear orthogonal coordinate system α,β,γ by $u_\alpha(\alpha,\beta,\gamma,t)$, $u_\beta(\alpha,\beta,\gamma,t)$, and $u_\gamma(\alpha,\beta,\gamma,t)$. Then the strain components will be denoted by $e_\alpha, e_\beta, e_\gamma, e_{\beta\gamma}, e_{\alpha\gamma}, e_{\alpha\beta}$. The symbols $e_\alpha, e_\beta, e_\gamma$ refer to the strains of elongation along three mutually perpendicular directions α, β, γ. $e_{\alpha\beta} = e_{\beta\alpha}$, $e_{\beta\gamma} = e_{\gamma\beta}$, $e_{\gamma\alpha} = e_{\alpha\gamma}$ are the shear strains (variations in angles) in three mutually perpendicular planes. The strain components of the ith layer are related to the displacements by the following equations:

$$\begin{aligned}
e^i_\alpha &= \frac{1}{H_1}\frac{\partial u^i_\alpha}{\partial\alpha} + \frac{1}{H_1H_2}\frac{\partial H_1}{\partial\beta}u^i_\beta + \frac{1}{H_1H_3}\frac{\partial H_1}{\partial\gamma}u^i_\gamma,\\
e^i_\beta &= \frac{1}{H_2}\frac{\partial u^i_\beta}{\partial\beta} + \frac{1}{H_2H_3}\frac{\partial H_2}{\partial\gamma}u^i_\gamma + \frac{1}{H_1H_2}\frac{\partial H_2}{\partial\alpha}u^i_\alpha,\\
e^i_\gamma &= \frac{1}{H_3}\frac{\partial u^i_\gamma}{\partial\gamma} + \frac{1}{H_3H_1}\frac{\partial H_3}{\partial\alpha}u^i_\alpha + \frac{1}{H_3H_2}\frac{\partial H_3}{\partial\beta}u^i_\beta,\\
e^i_{\alpha\beta} &= \frac{H_1}{H_2}\frac{\partial}{\partial\beta}\left(\frac{1}{H_1}u^i_\alpha\right) + \frac{H_2}{H_1}\frac{\partial}{\partial\alpha}\left(\frac{1}{H_2}u^i_\beta\right),\\
e^i_{\alpha\gamma} &= \frac{H_3}{H_1}\frac{\partial}{\partial\alpha}\left(\frac{1}{H_3}u^i_\gamma\right) + \frac{H_1}{H_3}\frac{\partial}{\partial\gamma}\left(\frac{1}{H_1}u^i_\alpha\right),\\
e^i_{\beta\gamma} &= \frac{H_2}{H_3}\frac{\partial}{\partial\gamma}\left(\frac{1}{H_2}u^i_\beta\right) + \frac{H_3}{H_2}\frac{\partial}{\partial\beta}\left(\frac{1}{H_3}u^i_\gamma\right).
\end{aligned} \tag{1.3.1}$$

In an orthogonal curvilinear coordinate system α,β,γ the stress state of an elastic body is characterized by the stresses σ_{ij} $(i,j=\alpha,\beta,\gamma)$, where $\sigma_\alpha, \sigma_\beta, \sigma_\gamma$ denote the normal stresses acting on the areas perpendicular to the coordinate axes α,β,γ. The other stress components are shear stresses acting on the indicated areas. In this case, in accordance with the law of equal complementary shear stresses, we have $\tau_{\alpha\beta} = \tau_{\beta\alpha}$, $\tau_{\beta\gamma} = \tau_{\gamma\beta}$, $\tau_{\alpha\gamma} = \tau_{\gamma\alpha}$. In the chosen system of curvilinear coordinates, α, β, γ, the partial differential equations for describing the equilibrium of the elastic medium (for the ith layer) have the following form (for the sake of simplicity the index i characteristic of the layer will be omitted):

$$\begin{aligned}
&\frac{\partial}{\partial \alpha}(H_2H_3\sigma_\alpha) + \frac{\partial}{\partial \beta}(H_1H_3\tau_{\alpha\beta}) + \frac{\partial}{\partial \gamma}(H_1H_2\tau_{\alpha\gamma}) - \sigma_\beta H_3\frac{\partial H_2}{\partial \alpha}\\
&\quad - \sigma_\gamma H_2\frac{\partial H_3}{\partial \alpha} + \tau_{\alpha\beta}H_3\frac{\partial H_1}{\partial \beta} + \tau_{\alpha\gamma}H_2\frac{\partial H_1}{\partial \gamma} + H_1H_2H_3F_\alpha = H_1H_2H_3,\\
&\frac{\partial}{\partial \beta}(H_1H_3\sigma_\beta) + \frac{\partial}{\partial \gamma}(H_1H_2\tau_{\beta\gamma}) + \frac{\partial}{\partial \alpha}(H_2H_3\tau_{\alpha\beta}) - \sigma_\alpha H_1\frac{\partial H_3}{\partial \beta}\\
&\quad - \sigma_\alpha H_3\frac{\partial H_1}{\partial \beta} + \tau_{\beta\gamma}H_1\frac{\partial H_2}{\partial \gamma} + \tau_{\alpha\beta}H_3\frac{\partial H_2}{\partial \alpha} + H_1H_2H_3F_\beta = H_1H_2H_3,\\
&\frac{\partial}{\partial \gamma}(H_1H_2\sigma_\gamma) + \frac{\partial}{\partial \alpha}(H_2H_3\tau_{\alpha\gamma}) + \frac{\partial}{\partial \beta}(H_1H_2\tau_{\beta\gamma}) - \sigma_\alpha H_2\frac{\partial H_1}{\partial \gamma}\\
&\quad - \sigma_\beta H_1\frac{\partial H_2}{\partial \gamma} + \tau_{\alpha\gamma}H_2\frac{\partial H_3}{\partial \alpha} + \tau_{\beta\gamma}H_1\frac{\partial H_3}{\partial \beta} + H_1H_2H_3F_\gamma = H_1H_2H_3,
\end{aligned} \tag{1.3.2}$$

where F_α, F_β and F_γ are the body forces.

After inertial terms have been added to the equilibrium Eq. (1.3.2) according to the principle of d'Alembert, they turn into the equations of motion:

$$\begin{aligned}
&\frac{\partial}{\partial \alpha}(H_2H_3\sigma_\alpha) + \frac{\partial}{\partial \beta}(H_1H_3\tau_{\alpha\beta}) + \frac{\partial}{\partial \gamma}(H_1H_2\tau_{\alpha\gamma}) - \sigma_\beta H_3\frac{\partial H_2}{\partial \alpha}\\
&\quad - \sigma_\gamma H_2\frac{\partial H_3}{\partial \alpha} + \tau_{\alpha\beta}H_3\frac{\partial H_1}{\partial \beta} + \tau_{\alpha\gamma}H_2\frac{\partial H_1}{\partial \gamma} + H_1H_2H_3F_\alpha = H_1H_2H_3\rho\frac{\partial^2 u_\alpha}{\partial t^2},\\
&\frac{\partial}{\partial \beta}(H_1H_3\sigma_\beta) + \frac{\partial}{\partial \gamma}(H_1H_2\tau_{\beta\gamma}) + \frac{\partial}{\partial \alpha}(H_2H_3\tau_{\alpha\beta}) - \sigma_\alpha H_1\frac{\partial H_3}{\partial \beta}\\
&\quad - \sigma_\alpha H_3\frac{\partial H_1}{\partial \beta} + \tau_{\beta\gamma}H_1\frac{\partial H_2}{\partial \gamma} + \tau_{\alpha\beta}H_3\frac{\partial H_2}{\partial \alpha} + H_1H_2H_3F_\beta = H_1H_2H_3\rho\frac{\partial^2 u_\beta}{\partial t^2},\\
&\frac{\partial}{\partial \gamma}(H_1H_2\sigma_\gamma) + \frac{\partial}{\partial \alpha}(H_2H_3\tau_{\alpha\gamma}) + \frac{\partial}{\partial \beta}(H_1H_2\tau_{\beta\gamma}) - \sigma_\alpha H_2\frac{\partial H_1}{\partial \gamma}\\
&\quad - \sigma_\beta H_1\frac{\partial H_2}{\partial \gamma} + \tau_{\alpha\gamma}H_2\frac{\partial H_3}{\partial \alpha} + \tau_{\beta\gamma}H_1\frac{\partial H_3}{\partial \beta} + H_1H_2H_3F_\gamma = H_1H_2H_3\rho\frac{\partial^2 u_\gamma}{\partial t^2},
\end{aligned} \tag{1.3.3}$$

where ρ is the mass density, and t refers to time.

Stress-strain relations are defined by the generalized Hooke's law for inhomogeneous anisotropic bodies:

$$
\begin{aligned}
\sigma_\alpha &= c_{11}e_\alpha + c_{12}e_\beta + c_{13}e_\gamma + c_{14}e_{\beta\gamma} + c_{15}e_{\alpha\gamma} + c_{16}e_{\alpha\beta},\\
\sigma_\beta &= c_{12}e_\alpha + c_{22}e_\beta + c_{23}e_\gamma + c_{24}e_{\beta\gamma} + c_{25}e_{\alpha\gamma} + c_{26}e_{\alpha\beta},\\
\sigma_\gamma &= c_{13}e_\alpha + c_{23}e_\beta + c_{33}e_\gamma + c_{34}e_{\beta\gamma} + c_{35}e_{\gamma\alpha} + c_{36}e_{\alpha\beta},\\
\sigma_{\beta\gamma} &= c_{14}e_\alpha + c_{24}e_\beta + c_{34}e_\gamma + c_{44}e_{\beta\gamma} + c_{45}e_{\gamma\alpha} + c_{46}e_{\alpha\beta},\\
\sigma_{\gamma\alpha} &= c_{15}e_\alpha + c_{25}e_\beta + c_{35}e_\gamma + c_{45}e_{\beta\gamma} + c_{55}e_{\gamma\alpha} + c_{56}e_{\alpha\beta},\\
\tau_{\alpha\beta} &= c_{16}e_\alpha + c_{26}e_\beta + c_{36}e_\gamma + c_{46}e_{\beta\gamma} + c_{56}e_{\alpha\gamma} + c_{66}e_{\alpha\beta},
\end{aligned}
\tag{1.3.4}
$$

or by inverse relations:

$$
\begin{aligned}
e_\alpha &= a_{11}\sigma_\alpha + a_{12}\sigma_\beta + a_{13}\sigma_\gamma + a_{14}\tau_{\beta\gamma} + a_{15}\tau_{\alpha\gamma} + a_{16}\tau_{\alpha\beta},\\
e_\beta &= a_{12}\sigma_\alpha + a_{22}\sigma_\beta + a_{23}\sigma_\gamma + a_{24}\tau_{\beta\gamma} + a_{25}\tau_{\alpha\gamma} + a_{26}\tau_{\alpha\beta},\\
e_\gamma &= a_{13}\sigma_\alpha + a_{23}\sigma_\beta + a_{33}\sigma_\gamma + a_{34}\tau_{\beta\gamma} + a_{35}\tau_{\gamma\alpha} + a_{36}\tau_{\alpha\beta},\\
e_{\beta\gamma} &= a_{14}\sigma_\alpha + a_{24}\sigma_\beta + a_{34}\sigma_\gamma + a_{44}\tau_{\beta\gamma} + a_{45}\tau_{\gamma\alpha} + a_{46}\tau_{\alpha\beta},\\
e_{\gamma\alpha} &= a_{15}\sigma_\alpha + a_{25}\sigma_\beta + a_{35}\sigma_\gamma + a_{45}\tau_{\beta\gamma} + a_{55}\tau_{\gamma\alpha} + a_{5}\tau_{\alpha\beta},\\
e_{\alpha\beta} &= a_{16}\sigma_\alpha + a_{26}\sigma_\beta + a_{36}\sigma_\gamma + a_{46}\tau_{\beta\gamma} + a_{56}\tau_{\alpha\gamma} + a_{66}\tau_{\alpha\beta},
\end{aligned}
\tag{1.3.5}
$$

where c_{ij} and a_{ij} are elastic constants (independent of a stress state).

For the general case of inhomogeneous elastic bodies, the elastic constants a_{ij} and c_{ij} as well as the density ρ entering the equations of motion (1.3.3) are specified functions of coordinates. We will assume that the values a_{ij} and c_{ij} as well as their derivatives up to the second order are continuous coordinate functions.

For different types of symmetry of elastic properties Hooke's law is specified in the same manner as for homogeneous bodies, i.e., given by Eqs. (1.3.4) and (1.3.5). In this case, the stress-strain relations differ from the corresponding relations for homogeneous bodies just in that the elasticity moduli and the compliance factors are now functions of coordinates.

If each point includes one plane of elastic symmetry tangential to the coordinate surface $\gamma = \text{const.}$, Hooke's law (1.3.5), which contains 13 independent elastic constants, will take the form:

$$
\begin{aligned}
e_\alpha &= a_{11}\sigma_\alpha + a_{12}\sigma_\beta + a_{13}\sigma_\gamma + a_{16}\tau_{\alpha\beta},\\
e_\beta &= a_{12}\sigma_\alpha + a_{22}\sigma_\beta + a_{23}\sigma_\gamma + a_{26}\tau_{\alpha\beta},\\
e_\gamma &= a_{13}\sigma_\alpha + a_{23}\sigma_\beta + a_{33}\sigma_\gamma + a_{36}\tau_{\alpha\beta},\\
2e_{\beta\gamma} &= a_{44}\tau_{\beta\gamma} + a_{45}\tau_{\alpha\beta},\\
2e_{\alpha\gamma} &= a_{45}\tau_{\beta\gamma} + a_{55}\tau_{\alpha\gamma},\\
2e_{\alpha\beta} &= a_{16}\sigma_\alpha + a_{26}\sigma_\beta + a_{36}\sigma_\gamma + a_{66}\tau_{\alpha\beta}.
\end{aligned}
\tag{1.3.6}
$$

In the case of an orthotropic material, when the principal directions of elasticity coincide with the directions of coordinate axes, the number of elastic coefficients decreases to nine. Then Hooke's law can be written in the following form (Sokolnikoff [25]):

$$\begin{aligned} e_\alpha &= a_{11}\sigma_\alpha + a_{12}\sigma_\beta + a_{13}\sigma_\gamma, \\ e_\beta &= a_{12}\sigma_\alpha + a_{22}\sigma_\beta + a_{23}\sigma_\gamma, \\ e_\gamma &= a_{13}\sigma_\alpha + a_{23}\sigma_\beta + a_{33}\sigma_\gamma, \\ 2e_{\beta\gamma} &= a_{44}\tau_{\beta\gamma}, \\ 2e_{\alpha\gamma} &= a_{55}\tau_{\alpha\gamma}, \\ 2e_{\alpha\beta} &= a_{66}\tau_{\alpha\beta}. \end{aligned} \tag{1.3.7}$$

The dependencies between the elastic constants c_{ij} and a_{ij} for the orthotropic material read:

$$\begin{aligned} & c_{11} = \frac{1}{\Omega}\left(a_{22}a_{33} - a_{23}^2\right), \quad c_{12} = c_{21} = \frac{1}{\Omega}\left(a_{31}a_{23} - a_{21}a_{33}\right) \\ & c_{13} = c_{31} = \frac{1}{\Omega}\left(a_{21}a_{32} - a_{31}a_{22}\right), \quad c_{22} = \frac{1}{\Omega}\left(a_{11}a_{33} - a_{13}^2\right) \\ & c_{23} = c_{32} = \frac{1}{\Omega}\left(a_{12}a_{31} - a_{11}a_{32}\right), \quad c_{33} = \frac{1}{\Omega}\left(a_{11}a_{22} - a_{12}^2\right) \\ & c_{44} = a_{44}, \quad c_{55} = a_{55}, \quad c_{55} = a_{55} \\ & \Omega = \left(a_{11}a_{22} - a_{12}^2\right)a_{66} + 2a_{12}a_{16}a_{26} - a_{11}a_{26}^2 - a_{22}a_{16}^2 \end{aligned} \tag{1.3.8}$$

Elastic constants for an orthotropic material can be expressed in terms of technical constants as follows:

$$\begin{aligned} & a_{11} = \frac{1}{E_\alpha}, \quad a_{22} = \frac{1}{E_\beta}, \quad a_{33} = \frac{1}{E_\gamma}, \\ & a_{12} = -\frac{\nu_{\beta\alpha}}{E_\alpha} = -\frac{\nu_{\alpha\beta}}{E_\beta}, \quad a_{13} = -\frac{\nu_{\alpha\gamma}}{E_\gamma} = -\frac{\nu_{\gamma\alpha}}{E_\alpha}, \quad a_{23} = -\frac{\nu_{\gamma\beta}}{E_\beta} = -\frac{\nu_{\beta\gamma}}{E_\gamma}, \\ & a_{44} = \frac{1}{G_{\beta\gamma}}, \quad a_{55} = \frac{1}{G_{\alpha\gamma}}, \quad a_{66} = \frac{1}{G_{\alpha\beta}}, \end{aligned} \tag{1.3.9}$$

where $E_\alpha, E_\beta, E_\gamma$ are the moduli of elasticity in the directions α, β, γ, respectively. $G_{\beta\gamma}, G_{\alpha\gamma}, G_{\alpha\beta}$ denote the shear moduli for planes parallel to coordinate surfaces $\alpha = \text{const.}$, $\beta = \text{const.}$, $\gamma = \text{const.}$, and $\nu_{\alpha\beta}, \nu_{\beta\alpha}, \nu_{\beta\gamma}, \nu_{\gamma\beta}, \nu_{\alpha\gamma}, \nu_{\gamma\alpha}$ are Poisson's ratios, which characterize transverse compression under tension in the direction of the coordinate axes (here the first index represents the direction of transverse compression and the second index indicates the force direction).

For a transversally-isotropic body containing, for example, the surface, $\gamma = \text{const.}$, on which all directions are equivalent with respect to the elastic properties, the number of independent elastic constants is equal to five. In this case we have:

$$\begin{aligned} e_\alpha &= a_{11}\sigma_\alpha + a_{12}\sigma_\beta + a_{13}\sigma_\gamma, \quad e_{\beta\gamma} = a_{44}\tau_{\beta\gamma} \\ e_\beta &= a_{12}\sigma_\alpha + a_{11}\sigma_\beta + a_{13}\sigma_\gamma, \quad e_{\alpha\gamma} = \tau_{\alpha\gamma} \\ e_\gamma &= a_{13}(\sigma_\alpha + \sigma_\beta) + a_{33}\sigma_\gamma, \quad e_{\alpha\beta} = 2(a_{11} - a_{12})\tau_{\alpha\beta} \end{aligned} \tag{1.3.10}$$

In the case of an isotropic body all directions are equivalent and any plane in the arbitrary point is a plane of elastic symmetry. Then the equations of the generalized Hooke's law read:

$$\begin{aligned} e_\alpha &= \frac{1}{E}\left[\sigma_\alpha - \nu\left(\sigma_\beta + \sigma_\gamma\right)\right], \quad e_{\beta\gamma} = \frac{1}{G}\tau_{\beta\gamma}, \\ e_\beta &= \frac{1}{E}\left[\sigma_\beta - \nu\left(\sigma_\alpha + \sigma_\gamma\right)\right], \quad e_{\gamma\alpha} = \frac{1}{G}\tau_{\gamma\alpha}, \\ e_\gamma &= \frac{1}{E}\left[\sigma_\gamma - \nu\left(\sigma_\alpha + \sigma_\beta\right)\right], \quad e_{\alpha\beta} = \frac{1}{G}\tau_{\alpha\beta}, \end{aligned} \tag{1.3.11}$$

where E is Young's modulus, ν is Poisson's ratio, $G = E/2(1+\nu)$ is the shear modulus. The number of independent elastic constants becomes equal to three.

1.4 Basic Relations for Classical Shell Models

1.4.1 Strains and Displacements of the Shell

The idea of the Kirchhoff-Love hypothesis (Love [17]) (classical shell model) lies in the fact that the elongation in the normal direction to the coordinate surface is of the order of one percent and that the transverse shears are assumed to be approximately equal to zero. Because the normal stresses at the areas parallel to the coordinate surface are rather small in comparison with the same stresses at the areas, which are perpendicular to the coordinate surface, they can be neglected. By using the straight-normal hypothesis, the problem of deformation of a laminated anisotropic shell can be reduced to the problem of deformation of an arbitrary coordinate surface. This makes it possible to solve a two-dimensional problem of elasticity theory instead of a three-dimensional one.

The use of the Kirchhoff-Love hypothesis makes it possible to express displacements of shell points that do not lie on the coordinate surface in terms of the components of displacements of the coordinate surface as follows:

$$\begin{aligned} u_\alpha(\alpha,\beta,\gamma) &= u(\alpha,\beta) + \gamma\vartheta_\alpha(\alpha,\beta), \\ u_\beta(\alpha,\beta,\gamma) &= v(\alpha,\beta) + \gamma\vartheta_\beta(\alpha,\beta), \\ u_\gamma(\alpha,\beta,\gamma) &= w(\alpha,\beta), \end{aligned} \tag{1.4.1}$$

where $u(\alpha, \beta), v(\alpha, \beta), w(\alpha, \beta)$ are the displacements of the coordinate surface in the directions α, β, γ, respectively, where the accuracy of the summand is of the second-order, ϑ_α and ϑ_β are the rotation angles of the coordinate surface in the planes $\alpha =$ const. and $\beta =$ const., respectively, by:

$$\vartheta_\alpha = -\frac{1}{A}\frac{\partial w}{\partial \alpha} + \frac{u}{R_\alpha}, \quad \vartheta_\beta = -\frac{1}{B}\frac{\partial w}{\partial \beta} + \frac{v}{R_\alpha}. \tag{1.4.2}$$

In this context, we suppose that the strains e_{ij} and the rotation of the shell element relative to the normal γ of the coordinate surface are small quantities of higher order when compared to the rotations relative to the α and β axes:

$$e_{\alpha\gamma} = e_{\beta\gamma} = e_\gamma = 0. \tag{1.4.3}$$

We assume that the displacements across the thickness (1.4.1) are linearly distributed (i.e., we consider thin-walled elastic bodies). Then the strains $e_\alpha, e_\beta, e_{\alpha\beta}$ can be represented by:

$$e_\alpha = \varepsilon_\alpha + \gamma\kappa_\alpha, \quad e_\beta = \varepsilon_\beta + \gamma\kappa_\beta, \quad e_{\alpha\beta} = \varepsilon_{\alpha\beta} + \gamma 2\kappa_{\alpha\beta}, \tag{1.4.4}$$

where $\varepsilon_\alpha, \varepsilon_\beta, \varepsilon_{\alpha\beta}$ and $\kappa_\alpha, \kappa_\beta, \kappa_{\alpha\beta}$ are the components of tangential and bending strains, respectively. These components can be expressed in terms of the displacements and rotation angles of the coordinate surface as follows:

$$\begin{aligned}
\varepsilon_\alpha &= \frac{1}{A}\frac{\partial u}{\partial \alpha} + \frac{1}{AB}\frac{\partial A}{\partial \beta}v + \frac{w}{R_\alpha},\\
\varepsilon_\beta &= \frac{1}{B}\frac{\partial u}{\partial \beta} + \frac{1}{AB}\frac{\partial B}{\partial \alpha}u + \frac{w}{R_\beta},\\
\varepsilon_{\alpha\beta} &= \frac{A}{B}\frac{\partial}{\partial \beta}\left(\frac{u}{A}\right) + \frac{B}{A}\frac{\partial}{\partial \alpha}\left(\frac{v}{B}\right),\\
\kappa_\alpha &= \frac{1}{A}\frac{\partial \vartheta_\alpha}{\partial \alpha} + \frac{1}{AB}\frac{\partial A}{\partial \beta}\vartheta_\beta,\\
\kappa_\beta &= \frac{1}{B}\frac{\partial \vartheta_\beta}{\partial \beta} + \frac{1}{AB}\frac{\partial B}{\partial \alpha}\vartheta_\alpha,\\
2\kappa_{\alpha\beta} &= \frac{1}{A}\frac{\partial \vartheta_\beta}{\partial \alpha} + \frac{1}{B}\frac{\partial \vartheta_\alpha}{\partial \beta} - \frac{1}{AB}\left(\frac{\partial A}{\partial \beta}\vartheta_\alpha + \frac{\partial B}{\partial \alpha}\vartheta_\beta\right)\\
&\quad + \frac{1}{R_\alpha}\left(\frac{1}{B}\frac{\partial u}{\partial \beta} - \frac{1}{AB}\frac{\partial B}{\partial \alpha}u\right) + \frac{1}{R_\beta}\left(\frac{1}{A}\frac{\partial v}{\partial \alpha} - \frac{1}{AB}\frac{\partial A}{\partial \beta}v\right).
\end{aligned} \tag{1.4.5}$$

By adopting the Kirchhoff-Love hypothesis for the shell package as a whole, we have to satisfy conditions of rigid contact of the layers in terms of displacements, namely:

$$u_{\alpha}^{i} = u_{\alpha}^{i+1}, \quad u_{\beta}^{i} = u_{\beta}^{i+1}, \quad u_{\gamma}^{i} = u_{\gamma}^{i+1} \quad \text{for} \quad \gamma = \gamma_i(\alpha, \beta) \quad (i = 1, 2, \ldots, n-1). \tag{1.4.6}$$

If we retain only linear terms in Eqs. (1.4.2) and (1.4.5) the relations of the linear theory of thin shells are obtained.

1.4.2 Equilibrium Equations

In order to reduce a three-dimensional problem to two-dimensional one that has already been carried out in the theory of thin laminated shells, the following integral characteristics—forces and moments—are introduced instead of the components of a stress tensor:

$$\begin{aligned}
N_{\alpha} &= \sum_{i=1}^{n} \int_{\gamma_{i-1}}^{\gamma_i} \sigma_{\alpha}\left(1 + \frac{\gamma}{R_{\beta}}\right)\mathrm{d}\gamma, \quad N_{\beta} = \sum_{i=1}^{n} \int_{\gamma_{i-1}}^{\gamma_i} \sigma_{\beta}\left(1 + \frac{\gamma}{R_{\alpha}}\right)\mathrm{d}\gamma, \\
N_{\alpha\beta} &= \sum_{i=1}^{n} \int_{\gamma_{i-1}}^{\gamma_i} \tau_{\alpha\beta}\left(1 + \frac{\gamma}{R_{\beta}}\right)\mathrm{d}\gamma, \quad N_{\beta\alpha} = \sum_{i=1}^{n} \int_{\gamma_{i-1}}^{\gamma_i} \tau_{\beta\alpha}\left(1 + \frac{\gamma}{R_{\alpha}}\right)\mathrm{d}\gamma, \\
M_{\alpha} &= \sum_{i=1}^{n} \int_{\gamma_{i-1}}^{\gamma_i} \sigma_{\alpha}\left(1 + \frac{\gamma}{R_{\beta}}\right)\gamma\mathrm{d}\gamma, \quad M_{\beta} = \sum_{i=1}^{n} \int_{\gamma_{i-1}}^{\gamma_i} \sigma_{\beta}\left(1 + \frac{\gamma}{R_{\alpha}}\right)\gamma\mathrm{d}\gamma, \\
M_{\alpha\beta} &= \sum_{i=1}^{n} \int_{\gamma_{i-1}}^{\gamma_i} \tau_{\alpha\beta}\left(1 + \frac{\gamma}{R_{\beta}}\right)\gamma\mathrm{d}\gamma, \quad M_{\beta\alpha} = \sum_{i=1}^{n} \int_{\gamma_{i-1}}^{\gamma_i} \tau_{\beta\alpha}\left(1 + \frac{\gamma}{R_{\alpha}}\right)\gamma\mathrm{d}\gamma.
\end{aligned} \tag{1.4.7}$$

The forces and moments $N_{\alpha}, N_{\alpha\beta}, M_{\alpha}, M_{\alpha\beta}$, referring to the unit length of the line $\alpha = \text{const.}$, are statically equivalent to the stresses that act in the normal section of the shell perpendicularly to the direction $\beta = \text{const.}$ (Fig. 1.4). The symbols $N_{\beta}, N_{\beta\alpha}, M_{\beta}, M_{\beta\alpha}$ have an analogous meaning. In this case, N_{α}, N_{β} are the normal forces, $N_{\alpha\beta}, N_{\beta\alpha}$ are the shearing forces, $M_{\alpha}, M_{\alpha\beta}$ are the bending and twisting moments at the section $\alpha = \text{const.}$, and $M_{\beta}, M_{\beta\alpha}$ are the bending and twisting moments at the section $\beta = \text{const.}$

Let us substitute the surface forces applied to the surfaces, which bound the shell, by statically equivalent forces, having in mind that by transferring the external forces on the coordinate surface, the additional moments can be neglected. At the same time, we may assume with a high accuracy that the coordinate surface is loaded only by forces distributed across. Then the vector characterizing the intensity of surface loads can be written as:

$$\bar{q} = \{q_a, q_\beta, q_\gamma\}. \tag{1.4.8}$$

Following Novozhilov [20], we put:

$$S = N_{\alpha\beta} - \frac{M_{\beta\alpha}}{R_\beta} = N_{\beta\alpha} - \frac{M_{\beta\alpha}}{R_\alpha}, \quad H = M_{\alpha\beta} = M_{\beta\alpha}. \tag{1.4.9}$$

The differential equilibrium equations for the element of the coordinate surface bounded by arcs of coordinate lines and acted upon by external forces, internal forces, and moments can be obtained from the Lagrange principle. According to this principle, the sum of all works performed by external forces on virtual displacements is equal to the work of internal forces on variations of strain components. The equilibrium equations have the form:

$$\begin{gathered}
\frac{\partial}{\partial\alpha}(BN_\alpha) - N_\beta\frac{\partial B}{\partial\alpha} + \frac{1}{A}\frac{\partial}{\partial\beta}(A^2S) + \frac{\partial}{\partial\beta}\left(\frac{A}{R_\alpha}H\right) + \frac{1}{R_\beta}\frac{\partial A}{\partial\beta}H + \frac{AB}{R_\alpha}Q_\alpha + ABq_\alpha = 0,\\
\frac{\partial}{\partial\beta}(AN_\beta) - N_\alpha\frac{\partial A}{\partial\beta} + \frac{1}{B}\frac{\partial}{\partial\alpha}(B^2S) + \frac{\partial}{\partial\alpha}\left(\frac{B}{R_\beta}H\right) + \frac{1}{R_\alpha}\frac{\partial B}{\partial\alpha}H + \frac{AB}{R_\beta}Q_\beta + ABq_\beta = 0,\\
\frac{\partial}{\partial\alpha}(BQ_\alpha) + \frac{\partial}{\partial\beta}(AQ_\beta) - \frac{AB}{R_\alpha}N_\alpha - \frac{AB}{R_\beta}N_\beta + ABq_\gamma = 0,\\
\frac{1}{A}\frac{\partial}{\partial\beta}(A^2H) + \frac{\partial}{\partial\alpha}(BM_\alpha) - \frac{\partial B}{\partial\alpha}M_\beta - ABQ_\alpha - AB\left(N_\alpha - \frac{1}{R_\beta}M_\beta\right)\vartheta_\alpha - ABS\vartheta_\beta = 0,\\
\frac{1}{B}\frac{\partial}{\partial\alpha}(B^2H) + \frac{\partial}{\partial\beta}(AM_\beta) - M_\alpha\frac{\partial A}{\partial\beta} - ABQ_\beta - AB\left(N_\beta - \frac{1}{R_\alpha}M_\alpha\right)\vartheta_\beta - ABS\vartheta_\alpha = 0.
\end{gathered} \tag{1.4.10}$$

These equations, referring to the system of coordinates α, β on the undeformed coordinate surface of the shell, are derived for the case of conservative loads (1.4.8). By neglecting nonlinear terms, we arrive at the equations of the linear theory of thin shells.

1.4.3 Elasticity Relationships

The connections between the internal forces, moments, and strain components are given by elasticity relationships. In shell theory these relationships are similar to Hooke's law in the theory of elasticity. Suggesting that each point of the shell has one plane of elastic symmetry, parallel to the coordinate surface, and allowing for the Kirchhoff-Love hypothesis, the elasticity relationships can be written (Ambartsumyan [2, 3], Burton and Noor [5], Librescu [15], Reddy [22]) as follows:

$$\begin{aligned}
N_\alpha &= C_{11}\varepsilon_\alpha + C_{12}\varepsilon_\beta + C_{16}\varepsilon_{\alpha\beta} + K_{11}\kappa_\alpha + K_{12}\kappa_\beta + K_{16}\kappa_{\alpha\beta},\\
N_\beta &= C_{12}\varepsilon_\alpha + C_{22}\varepsilon_\beta + C_{26}\varepsilon_{\alpha\beta} + K_{12}\kappa_\alpha + K_{22}\kappa_\beta + K_{26}\kappa_{\alpha\beta},\\
S &= C_{16}\varepsilon_\alpha + C_{26}\varepsilon_\beta + C_{66}\varepsilon_{\alpha\beta} + K_{16}\kappa_\alpha + K_{26}\kappa_\beta + K_{66}\kappa_{\alpha\beta},\\
M_\alpha &= K_{11}\varepsilon_\alpha + K_{12}\varepsilon_\beta + K_{16}\varepsilon_{\alpha\beta} + D_{11}\kappa_\alpha + D_{12}\kappa_\beta + D_{16}\kappa_{\alpha\beta},\\
M_\beta &= K_{12}\varepsilon_\alpha + K_{22}\varepsilon_\beta + K_{26}\varepsilon_{\alpha\beta} + D_{12}\kappa_\alpha + D_{22}\kappa_\beta + D_{26}\kappa_{\alpha\beta},\\
H &= K_{16}\varepsilon_\alpha + K_{26}\varepsilon_\beta + K_{66}\varepsilon_{\alpha\beta} + D_{16}\kappa_\alpha + D_{26}\kappa_\beta + D_{66}\kappa_{\alpha\beta}.
\end{aligned} \tag{1.4.11}$$

Here the stiffness characteristics C_{mp}, K_{mp}, D_{mp} of the shells depend on the mechanical parameters and on the layer thickness. They are defined by:

$$\begin{aligned}
C_{mp} &= \sum_{i=1}^{n} \int_{\gamma_{i-1}}^{\gamma_i} B^i_{mp} \mathrm{d}\gamma, \quad K_{mp} = \sum_{i=1}^{n} \int_{\gamma_{i-1}}^{\gamma_i} B^i_{mp} \gamma \mathrm{d}\gamma,\\
D_{mp} &= \sum_{i=1}^{n} \int_{\gamma_{i-1}}^{\gamma_i} B^i_{mp} \gamma^2 \mathrm{d}\gamma \quad (m, p = 1, 2, 6),
\end{aligned} \tag{1.4.12}$$

where the functions B^i_{mp} are expressed in terms of mechanical characteristics of the ith layer as follows:

$$\begin{aligned}
B^i_{11} &= \frac{1}{\Omega_i}\left[a^i_{22}a^i_{66} - \left(a^i_{26}\right)^2\right], \quad B^i_{12} = \frac{1}{\Omega_i}\left[a^i_{16}a^i_{26} - a^i_{12}a^i_{66}\right],\\
B^i_{22} &= \frac{1}{\Omega_i}\left[a^i_{11}a^i_{66} - \left(a^i_{16}\right)^2\right], \quad B^i_{16} = \frac{1}{\Omega_i}\left[a^i_{12}a^i_{26} - a^i_{22}a^i_{16}\right],\\
B^i_{26} &= \frac{1}{\Omega_i}\left[a^i_{12}a^i_{16} - a^i_{11}a^i_{26}\right], \quad B^i_{66} = \frac{1}{\Omega_i}\left[a^i_{11}a^i_{22} - \left(a^i_{12}\right)^2\right],\\
\Omega_i &= \left[a^i_{11}a^i_{22} - \left(a^i_{12}\right)^2\right]a^i_{66} + 2a^i_{12}a^i_{16}a^i_{26} - a^i_{11}\left(a^i_{26}\right)^2 - a^i_{22}\left(a^i_{16}\right)^2.
\end{aligned} \tag{1.4.13}$$

For an inhomogeneous material the quantities a^i_{mp} (m, p = 1, 2, 6) are functions of coordinates and have the following form:

$$\begin{aligned}
a^i_{11} &= \frac{1}{E^i_\alpha}, \quad a^i_{22} = \frac{1}{E^i_\beta}, \quad a^i_{66} = \frac{1}{G^i_{\alpha\beta}}\\
a^i_{12} &= -\frac{\nu^i_{\beta\alpha}}{E^i_\alpha} = -\frac{\nu^i_{\alpha\beta}}{E^i_\beta}\\
a^i_{16} &= \frac{\eta^i_{\alpha\beta,\alpha}}{E^i_\alpha} = \frac{\eta^i_{\alpha,\alpha\beta}}{G^i_{\alpha\beta}}, \quad a^i_{26} = \frac{\eta^i_{\alpha\beta,\beta}}{E^i_\beta} = \frac{\eta^i_{\beta,\alpha\beta}}{G^i_{\alpha\beta}}
\end{aligned} \tag{1.5.14}$$

where E^i_α, E^i_β are the elasticity moduli in the α and β directions, respectively. $G^i_{\alpha\beta}$ is the shear modulus in the plane parallel to the coordinate surface, $v^i_{\alpha\beta}$ and $v^i_{\beta\alpha}$ are Poisson's ratios that characterize transverse compression (tension) in the direction of coordinate axes (the first index stands for the direction of the transverse compression, the second index for the direction of the force action). $\eta^i_{\alpha\beta,\alpha}, \eta^i_{\alpha\beta,\beta}$ are the coefficients of mutual influence that characterize shears in the coordinate surface caused by normal stresses in α and β directions. $\eta^i_{\alpha,\alpha\beta}$, $\eta^i_{\beta,\alpha\beta}$ are coefficients that characterize extensions caused by tangential stresses.

Let us consider the notation of elasticity relations for a number of cases that are important from an application's point-of-view, when the material and structure of the shell are anisotropic across the thickness.

For a single-layer isotropic shell of variable thickness $h(\alpha, \beta)$ the mechanical characteristics are:

$$E_\varepsilon = E_\beta = E, \quad v_{\alpha\beta} = v_{\beta\alpha} = v, \quad G_{\alpha\beta} = \frac{E}{2(1+v)} \tag{1.4.15}$$

As coordinate surface, we choose the median surface, i.e., the surface which is equidistant from the surfaces bounding the shell. Then the values C_{mp}, K_{mp}, D_{mp} defined by Eq. (1.4.12) become:

$$\begin{aligned} &C_{11} = C_{22} = \frac{Eh}{1-v^2}, \quad C_{12} = vC_{11}, \quad C_{66} = \frac{Eh}{2(1+v)}, \\ &D_{11} = D_{22} = \frac{Eh^2}{12(1-v^2)}, \quad D_{12} = vD_{11}, \quad D_{66} = \frac{Eh^3}{24(1+v)}, \\ &C_{16} = C_{26} = D_{16} = D_{26} = K_{11} = K_{12} = K_{22} = K_{16} = K_{26} = K_{66} = 0. \end{aligned} \tag{1.4.16}$$

The elasticity relations can be written as follows:

$$\begin{aligned} &N_\alpha = \frac{Eh}{1-v^2}(\varepsilon_\alpha + v\varepsilon_\beta), \quad N_\beta = \frac{Eh}{1-v^2}(v\varepsilon_\alpha + \varepsilon_\beta), \quad S = \frac{Eh}{2(1+v)}\varepsilon_{\alpha\beta}, \\ &M_\alpha = \frac{Eh^3}{12(1-v^2)}(\kappa_\alpha + v\kappa_\beta), \quad M_\beta = \frac{Eh^3}{12(1-v^2)}(v\kappa_\alpha + \kappa_\beta), \\ &H = \frac{Eh^2}{12(1+v)}\kappa_{\alpha\beta}. \end{aligned} \tag{1.4.17}$$

Let us consider a single-layered shell made of orthotropic material so that all three principal elasticity directions at each point coincide with the directions of the corresponding coordinate lines. Having chosen the median surface as a coordinate, we obtain:

$$
\begin{aligned}
&C_{11} = \frac{E_\alpha h}{1 - \nu_{\beta\alpha}\nu_{\alpha\beta}}, \quad C_{12} = \nu_{\alpha\beta} C_{11}, \\
&C_{22} = \frac{E_\beta h}{1 - \nu_{\beta\alpha}\nu_{\alpha\beta}}, \quad C_{66} = \nu_{\alpha\beta} G_{\alpha\beta} h, \\
&D_{11} = \frac{E_\alpha h^3}{12(1 - \nu_{\beta\alpha}\nu_{\alpha\beta})}, \quad D_{12} = \nu_{\alpha\beta} D_{11}, \\
&D_{22} = \frac{E_\alpha h^3}{12(1 - \nu_{\beta\alpha}\nu_{\alpha\beta})}, \quad D_{66} = G_{\alpha\beta}\frac{h^3}{12}, \\
&C_{16} = C_{26} = D_{16} = D_{26} = K_{11} = K_{12} = K_{22} = K_{16} = K_{26} = K_{66} = 0.
\end{aligned}
\tag{1.4.18}
$$

In this case, the elasticity relations become:

$$
\begin{aligned}
N_\alpha &= \frac{E_\alpha h}{1 - \nu_{\beta\alpha}\nu_{\alpha\beta}}\left(\varepsilon_\alpha + \nu_{\alpha\beta}\varepsilon_\beta\right), \quad N_\beta = \frac{E_\beta h}{1 - \nu_{\beta\alpha}\nu_{\alpha\beta}}\left(\nu_{\beta\alpha}\varepsilon_\alpha + \varepsilon_\beta\right), \\
S &= G_{\alpha\beta} h \varepsilon_{\alpha\beta}, \\
M_\alpha &= \frac{E_\alpha h^3}{12(1 - \nu_{\beta\alpha}\nu_{\alpha\beta})}\left(\kappa_\alpha + \nu_{\alpha\beta}\kappa_\beta\right), \quad M_\beta = \frac{E_\beta h^3}{12(1 - \nu_{\beta\alpha}\nu_{\alpha\beta})}\left(\nu_{\beta\alpha}\kappa_\alpha + \kappa_\beta\right), \\
H &= \frac{h^3}{6} G_{\alpha\beta}\,\kappa_{\alpha\beta}.
\end{aligned}
\tag{1.4.19}
$$

If the shell under consideration consists of an odd number of orthotropic or isotropic layers, located symmetrically with respect to the median surface, then, having it chosen as a coordinate surface, we obtain:

$$
\begin{aligned}
&N_\alpha = C_{11}\varepsilon_\alpha + C_{12}\varepsilon_\beta, \quad N_\beta = C_{12}\varepsilon_\alpha + C_{22}\varepsilon_\beta, \quad S = C_{66}\varepsilon_{\alpha\beta}, \\
&M_\alpha = D_{11}\kappa_\alpha + D_{12}\kappa_\beta, \quad M_\beta = D_{12}\kappa_\alpha + D_{22}\kappa_\beta, \quad H = 2D_{66}\kappa_{\alpha\beta}.
\end{aligned}
\tag{1.4.20}
$$

If the structure of the shell package is nonsymmetrical relatively to the median surface, then its choice as a coordinate surface does not result in a simplification of the elasticity relations. In this case, when choosing the coordinate surface, it is necessary to follow other path: When designing the sandwich-type variable-thickness shells of revolution manufactured by winding the tape of unidirectional material, it is convenient to choose the internal surface of a shell as coordinate one.

For shells composed of arbitrary number of orthotropic and isotropic layers, the elasticity relations in the case of arbitrary choice of a coordinate surface have the form:

$$\begin{aligned}
N_\alpha &= C_{11}\varepsilon_\alpha + C_{12}\varepsilon_\beta + K_{11}\kappa_\alpha + K_{12}\kappa_\beta,\\
N_\beta &= C_{12}\varepsilon_\alpha + C_{22}\varepsilon_\beta + K_{12}\kappa_\alpha + K_{22}\kappa_\beta,\\
S &= C_{66}\varepsilon_{\alpha\beta} + 2K_{66}\kappa_{\alpha\beta},\\
M_\alpha &= K_{11}\varepsilon_\alpha + K_{12}\varepsilon_\beta + D_{11}\kappa_\alpha + D_{12}\kappa_\beta,\\
M_\beta &= K_{12}\varepsilon_\alpha + K_{22}\varepsilon_\beta + D_{12}\kappa_\alpha + D_{22}\kappa_\beta,\\
H &= K_{66}\varepsilon_{\alpha\beta} + 2D_{66}\kappa_{\alpha\beta}.
\end{aligned} \tag{1.4.21}$$

As can be seen from Eq. (1.4.12), the coefficients $C_{11}, C_{12}, C_{22}, C_{66}$ are independent of the order the layers followed in the shell package, while the coefficients K_{mp} and D_{mp} depend on the choice of a coordinate surface and structure of the package of layers.

If the principal elasticity directions do not coincide with the coordinate lines and if they are rotated about the γ-axis by the angle ψ, then the values B^i_{mp} used in (1.4.11) in the coordinate system α, β are related to the same values $B^{(0)i}_{mp}$ in the coordinate system of principal elasticity directions as follows:

$$\begin{aligned}
B^i_{11} &= B^{(0)i}_{11}\cos^4\psi + 2\left(B^{(0)i}_{12} + 2B^{(0)i}_{66}\right)\sin^2\psi\cos^2\psi + B^{(0)i}_{22}\sin^4\psi\\
&\quad + 2\left(B^{(0)i}_{16}\cos^2\psi + B^{(0)i}_{26}\sin^2\psi\right)\sin 2\psi,\\
B^i_{22} &= B^{(0)i}_{11}\sin^4\psi + 2\left(B^{(0)i}_{12} + 2B^{(0)i}_{66}\right)\sin^2\psi\cos^2\psi + B^{(0)i}_{22}\cos^4\psi\\
&\quad - 2\left(B^{(0)i}_{16}\sin^2\psi + B^{(0)i}_{26}\cos^2\psi\right)\sin 2\psi,\\
B^i_{12} &= B^{(0)i}_{12} + \left[B^{(0)i}_{11} + B^{(0)i}_{22} - 2\left(B^{(0)i}_{12} + 2B^{(0)i}_{66}\right)\right]\sin^2\psi\cos^2\psi\\
&\quad + \left(B^{(0)i}_{26} - B^{(0)i}_{16}\right)\cos 2\psi\sin 2\psi,\\
B^i_{66} &= B^{(0)i}_{66} + \left[B^{(0)i}_{11} + B^{(0)i}_{22} - 2\left(B^{(0)i}_{12} + 2B^{(0)i}_{66}\right)\right]\sin^2\psi\cos^2\psi\\
&\quad + \left(B^{(0)i}_{26} - B^{(0)i}_{16}\right)\cos 2\psi\sin 2\psi,\\
B^i_{16} &= \left[B^{(0)i}_{22}\sin^2\psi - B^{(0)i}_{11}\cos^2\psi + \left(B^{(0)i}_{12} + 2B^{(0)i}_{66}\right)\cos 2\psi\right]\sin\psi\cos\psi\\
&\quad + B^{(0)i}_{16}\cos^2\psi\left(\cos^2\psi - 3\sin^2\psi\right) + B^{(0)i}_{26}\sin^2\psi\left(3\cos^2\psi - \sin^2\psi\right),\\
B^i_{26} &= \left[B^{(0)i}_{22}\cos^2\psi - B^{(0)i}_{11}\sin^2\psi - \left(B^{(0)i}_{12} + 2B^{(0)i}_{66}\right)\cos 2\psi\right]\sin\psi\cos\psi\\
&\quad + B^{(0)i}_{16}\sin^2\psi\left(3\cos^2\psi - \sin^2\psi\right) + B^{(0)i}_{26}\cos^2\psi\left(\cos^2\psi - 3\sin^2\psi\right).
\end{aligned} \tag{1.4.22}$$

The stresses arising under deformation of the ith layer can be expressed in terms of strain components for the coordinate surface by:

$$\begin{aligned}
\sigma_\alpha^i &= B_{11}^i\varepsilon_\alpha + B_{12}^i\varepsilon_\beta + B_{16}^i\varepsilon_{\alpha\beta} + \gamma\left(B_{11}^i\kappa_\alpha + B_{12}^i\kappa_\beta + 2B_{16}^i\kappa_{\alpha\beta}\right),\\
\sigma_\beta^i &= B_{12}^i\varepsilon_\alpha + B_{22}^i\varepsilon_\beta + B_{26}^i\varepsilon_{\alpha\beta} + \gamma\left(B_{12}^i\kappa_\alpha + B_{22}^i\kappa_\beta + 2B_{26}^i\kappa_{\alpha\beta}\right),\\
\tau_{\alpha\beta}^i &= B_{16}^i\varepsilon_\alpha + B_{26}^i\varepsilon_\beta + B_{66}^i\varepsilon_{\alpha\beta} + \gamma\left(B_{16}^i\kappa_\alpha + B_{26}^i\kappa_\beta + 2B_{66}^i\kappa_{\alpha\beta}\right).
\end{aligned} \tag{1.4.23}$$

In here the values B_{mp}^i are defined by (1.4.13).

1.4.4 Boundary Conditions

The kinematic relations between the strains expressed in terms of displacements (1.4.5), the equilibrium Eq. (1.4.10), and the elasticity relations (1.4.11) form a closed system of differential equations and can be used for describing the stress-strain state in the shell. However, in order to eliminate the arbitrariness in the solution of this system it is necessary to know boundary conditions. According to the order of this system of differential equations, four boundary conditions should be specified at all shell contours. Since the objects being considered are shells whose contours coincide with the lines of principal curvatures of a coordinate surface, static boundary conditions can be formulated by combinations of the following:

At the contour $\alpha = \text{const.}$:

$$N_\alpha, M_\alpha, \hat{S}_\alpha = S + \frac{2}{R_\beta}H, \quad \hat{Q}_\alpha = Q_\alpha + \frac{1}{B}\frac{\partial H}{\partial \beta}, \tag{1.4.24}$$

and at the contour $\beta = \text{const.}$:

$$N_\beta, M_\beta, \hat{S}_\beta = S + \frac{2}{R_\alpha}H, \quad \hat{Q}_\beta = Q_\beta + \frac{1}{A}\frac{\partial H}{\partial \alpha} \tag{1.4.25}$$

The quantities $\hat{S}_\alpha, \hat{S}_\beta$ and $\hat{Q}_\alpha, \hat{Q}_\beta$ are reduced shear and transverse forces.

Kinematic boundary conditions may be formulated in terms of displacements by combination of quantities:

At the contour $\alpha = \text{const.}$ from:

$$u, v, w, \vartheta_\alpha, \tag{1.4.26}$$

and at the contour $\beta = \text{const.}$ from:

$$u, v, w, \vartheta_\beta. \tag{1.4.27}$$

The boundary conditions can also be specified in mixed form, i.e., as combinations of forces and moments, strains and displacements.

1.5 Basic Relations for Refined Shell Models

In what follows we shall consider shells composed of anisotropic layers of variable thickness, connected in a single stack without sliding and separation. We assume that a plane of elastic symmetry parallel to a plane tangent to the coordinate surface exists. We will apply the straight line hypothesis for the entire shell layers across the thickness. The idea of this hypothesis is that an element that was initially normal to a coordinate (reference) surface remains rectilinear but not perpendicular to the deformed coordinate surface. In this case, it is supposed that the contraction over the stack thickness is absent and that the length of the element remains unchanged. Then, according to the above hypothesis, we arrive at a linear law for the displacement distribution across the thickness:

$$\begin{aligned} u_\alpha(\alpha,\beta,\gamma) &= u(\alpha,\beta) + \gamma\psi_\alpha(\alpha,\beta), \\ u_\beta(\alpha,\beta,\gamma) &= v(\alpha,\beta) + \gamma\psi_\beta(\alpha,\beta), \\ u_\gamma(\alpha,\beta,\gamma) &= w(\alpha,\beta), \end{aligned} \tag{1.5.1}$$

which satisfies conditions of perfect contact of adjacent layers. u, v, w denote the displacements of points of the coordinate surface in the directions α, β, γ respectively. Moreover, ψ_α, ψ_β are the rotation angles of the rectilinear element.

The kinematic hypothesis (1.5.1) is supplemented by a static hypothesis, which makes it possible to neglect normal transverse stresses σ_γ^i in the generalized Hooke's law relations when compared to the other stresses.

If the law for the distribution of displacements across the shell thickness (1.5.1) is substituted into Eq. (1.5.4), and if only linear terms for γ in the expressions for the strains are retained, we obtain:

$$\begin{aligned} e_\alpha(\alpha,\beta,\gamma) &= \varepsilon_\alpha(\alpha,\beta) + \gamma\kappa_\alpha(\alpha,\beta), \\ e_\beta(\alpha,\beta,\gamma) &= \varepsilon_\beta(\alpha,\beta) + \gamma\kappa_\beta(\alpha,\beta), \\ e_{\alpha\beta}(\alpha,\beta,\gamma) &= \varepsilon_{\alpha\beta}(\alpha,\beta) + 2\gamma\kappa_{\alpha\beta}(\alpha,\beta), \\ e_{\alpha\gamma}(\alpha,\beta,\gamma) &= \gamma_\alpha(\alpha,\beta), \quad e_{\beta\gamma}(\alpha,\beta,\gamma) = \gamma_\beta(\alpha,\beta), \end{aligned} \tag{1.5.2}$$

where:

$$
\begin{aligned}
\varepsilon_\alpha &= \frac{1}{A}\frac{\partial u}{\partial \alpha} + \frac{1}{AB}\frac{\partial A}{\partial \beta}v + k_1 w, \quad \varepsilon_\beta = \frac{1}{B}\frac{\partial v}{\partial \beta} + \frac{1}{AB}\frac{\partial B}{\partial \alpha}u + k_2 w,\\
\varepsilon_{\alpha\beta} &= \frac{A}{B}\frac{\partial}{\partial \beta}\left(\frac{u}{A}\right) + \frac{B}{A}\frac{\partial}{\partial \alpha}\left(\frac{v}{B}\right), \quad \kappa_\alpha = \frac{1}{A}\frac{\partial \psi_a}{\partial \alpha} + \frac{1}{AB}\frac{\partial A}{\partial \beta}\psi_\beta - k_1 \varepsilon_\alpha,\\
\kappa_\beta &= \frac{1}{B}\frac{\partial \psi_\beta}{\partial \beta} + \frac{1}{AB}\frac{\partial B}{\partial \alpha}\psi_\alpha - k_2 \varepsilon_\beta,\\
2\kappa_{\alpha\beta} &= \frac{A}{B}\frac{\partial}{\partial \beta}\left(\frac{\psi_\alpha}{A}\right) + \frac{B}{A}\frac{\partial}{\partial \alpha}\left(\frac{\psi_\beta}{B}\right) + \frac{k_1}{B}\left(\frac{\partial u}{\partial \beta} - \frac{u}{A}\frac{\partial B}{\partial \alpha}\right) + \frac{k_2}{A}\left(\frac{\partial v}{\partial \alpha} - \frac{v}{B}\frac{\partial A}{\partial \beta}\right),\\
\gamma_\alpha &= \psi_\alpha - \vartheta_\alpha, \quad \gamma_\beta = \psi_\beta - \vartheta_\beta,\\
\vartheta_\alpha &= -\frac{1}{A}\frac{\partial w}{\partial \alpha} + k_1 u, \quad \vartheta_\beta = -\frac{1}{B}\frac{\partial w}{\partial \beta} + k_2 v.
\end{aligned}
\tag{1.5.3}
$$

The symbols appearing in Eq. (1.5.3) have the following geometrical meaning: ε_α, ε_β characterize tension (compression) of the coordinate surface, $\varepsilon_{\alpha\beta}$ are shears, κ_α, κ_β, $\kappa_{\alpha\beta}$ are bending and twisting strains, and γ_α, γ_β are the rotation angles of the straight-line element attributed to transverse shears.

If we recall the basic idea according to which the three-dimensional elasticity problem is reduced to a two-dimensional problem for the deformation of the coordinate surface, we can replace the stresses by equivalent values of forces and moments:

$$
\begin{aligned}
N_\alpha &= \sum_{i=1}^{N}\int_{\gamma_{i-1}}^{\gamma_i} \sigma^i_\alpha(1+k_2\gamma)\mathrm{d}\gamma, \quad N_\beta = \sum_{i=1}^{N}\int_{\gamma_{i-1}}^{\gamma_i} \sigma^i_\beta(1+k_1\gamma)\mathrm{d}\gamma, \quad N_{\alpha\beta} = \sum_{i=1}^{N}\int_{\gamma_{i-1}}^{\gamma_i} \tau^i_{\alpha\beta}(1+k_2\gamma)\mathrm{d}\gamma\\
N_{\beta\alpha} &= \sum_{i=1}^{N}\int_{\gamma_{i-1}}^{\gamma_i} \tau^i_{\beta\alpha}(1+k_1\gamma)\mathrm{d}\gamma \quad Q_\alpha = \sum_{i=1}^{N}\int_{\gamma_{i-1}}^{\gamma_i} \tau^i_{\alpha\gamma}(1+k_2\gamma)\mathrm{d}\gamma \quad Q_\beta = \sum_{i=1}^{N}\int_{\gamma_{i-1}}^{\gamma_i} \tau^i_{\beta\gamma}(1+k_1\gamma)\mathrm{d}\gamma\\
M_\alpha &= \sum_{i=1}^{N}\int_{\gamma_{i-1}}^{\gamma_i} \sigma^i_\alpha(1+k_2\gamma)\gamma\mathrm{d}\gamma \quad M_\beta = \sum_{i=1}^{N}\int_{\gamma_{i-1}}^{\gamma_i} \sigma^i_\beta(1+k_1\gamma)\gamma\mathrm{d}\gamma\\
M_{\alpha\beta} &= \sum_{i=1}^{N}\int_{\gamma_{i-1}}^{\gamma_i} \tau^i_{\alpha\beta}(1+k_2\gamma)\gamma\mathrm{d}\gamma \quad M_{\beta\alpha} = \sum_{i=1}^{N}\int_{\gamma_{i-1}}^{\gamma_i} \tau^i_{\beta\alpha}(1+k_1\gamma)\gamma\mathrm{d}\gamma
\end{aligned}
\tag{1.5.4}
$$

where N_α, N_β are normal tangential forces, $N_{\alpha\beta}, N_{\beta\alpha}$ are shearing forces, Q_α, Q_β are shearing forces, M_α, M_β are bending moments, and $M_{\alpha\beta}, M_{\beta\alpha}$ are twisting moments.

The equilibrium equations of the element of the coordinate surface that have been derived allowing for variations in the shell thickness are:

$$
\begin{aligned}
&\frac{\partial}{\partial\alpha}(BN_\alpha)-\frac{\partial B}{\partial\alpha}N_\beta+\frac{\partial}{\partial\beta}(AN_{\beta\alpha})+\frac{\partial A}{\partial\beta}N_{\alpha\beta}+ABk_1Q_\alpha+ABq_\alpha=0,\\
&\frac{\partial}{\partial\beta}(AN_\beta)-\frac{\partial A}{\partial\beta}N_\alpha+\frac{\partial}{\partial\alpha}(BN_{\alpha\beta})+\frac{\partial B}{\partial\alpha}N_{\beta\alpha}+ABk_2Q_\beta+ABq_\beta=0,\\
&\frac{\partial}{\partial\alpha}(BQ_\alpha)+\frac{\partial}{\partial\beta}(AQ_\beta)-ABk_1N_\alpha-ABk_2N_\beta+ABq_\gamma=0,\\
&\frac{\partial}{\partial\alpha}(BM_\alpha)-\frac{\partial B}{\partial\alpha}M_\beta+\frac{\partial}{\partial\beta}(AM_{\beta\alpha})+\frac{\partial A}{\partial\beta}M_{\alpha\beta}-ABQ_\alpha+ABm_\alpha=0,\\
&\frac{\partial}{\partial\beta}(AM_\beta)-\frac{\partial A}{\partial\beta}M_\alpha+\frac{\partial}{\partial\alpha}(BM_{\alpha\beta})+\frac{\partial B}{\partial\alpha}M_{\beta\alpha}-ABQ_\beta+ABm_\beta=0,\\
&N_{\alpha\beta}-k_2M_{\beta\alpha}-N_{\beta\alpha}+k_1M_{\alpha\beta}=0.
\end{aligned}
\tag{1.5.5}
$$

The quantities $q_\alpha, q_\beta, q_\gamma, m_\alpha, m_\beta$ are determined by loads applied to the shell surface.

The relations of elasticity for orthotropic shells with symmetric structure across the thickness with respect to the chosen surface take the form:

$$
\begin{aligned}
N_\alpha&=C_{11}\varepsilon_\alpha+C_{12}\varepsilon_\beta, \quad N_\beta=C_{12}\varepsilon_\alpha+C_{22}\varepsilon_\beta,\\
N_{\alpha\beta}&=C_{66}\varepsilon_{\alpha\beta}+2k_2D_{66}\kappa_{\alpha\beta}, \quad N_{\beta\alpha}=C_{66}\varepsilon_{\alpha\beta}+2k_1D_{66}\kappa_{\alpha\beta},\\
M_\alpha&=D_{11}\kappa_\alpha+D_{12}\kappa_\beta, \quad M_\beta=D_{12}\kappa_\alpha+D_{22}\kappa_\beta,\\
M_{\beta\alpha}&=M_{\alpha\beta}=2D_{66}\kappa_{\alpha\beta}, \quad Q_\alpha=K_1\gamma_\alpha, Q_\beta=K_2\gamma_\beta,
\end{aligned}
\tag{1.5.6}
$$

where:

$$
\begin{aligned}
C_{11}&=\frac{E_\alpha h}{1-\nu_\alpha\nu_\beta}, \quad C_{12}=\nu_\beta C_{11}, \quad C_{22}=\frac{E_\beta h}{1-\nu_\alpha\nu_\beta}, C_{66}=G_{\alpha\beta}h,\\
D_{11}&=\frac{E_\alpha h^3}{12(1-\nu_\alpha\nu_\beta)}, \quad D_{12}=\nu_\beta D_{11}, \quad D_{22}=\frac{E_\beta h^3}{12(1-\nu_\alpha\nu_\beta)},\\
D_{66}&=\frac{G_{\alpha\beta}h^3}{12}, \quad K_1=\frac{5}{6}hG_{\alpha\gamma}, \quad K_2=\frac{5}{6}hG_{\beta\gamma}.
\end{aligned}
\tag{1.5.7}
$$

Here E_α, E_β, ν_α, ν_β are elastic moduli and Poisson's ratios in the α and β directions, $G_{\alpha\beta}$, $G_{\alpha\gamma}$, $G_{\beta\gamma}$ are shear moduli, and $h=h(x,y)$ is the shell thickness.

In order to solve the problem, it is necessary to complement the equations of the refined model by boundary conditions. To this end, five conditions at the contour of the coordinate surface should be formulated. Analogously to the classical model, the number of boundary conditions of the refined model must match the corresponding order of equations. Because the number of equations differs from the classical case new variants of boundary conditions must be formulated. The concrete statement depends on how the shell contours are fixed. In a general case, it is formulated in terms of forces, moments, displacements of coordinate surface, and rotation angles.

References

1. Altenbach H, Eremeyev A (2011) Shell-like structures: non-classical theories and application. Springer, Berlin
2. Ambartsumyan SA (1990) Fragments of the theory of anisotropic shells. World Scientific, Singapore
3. Ambartsumian SA (1991) Theory of anisotropic plates: strength, stability, vibration. Technomic, Lancaster
4. Ambartsumyan SA (1964) Theory of anisotropic shells. NASA Technical Translation
5. Burton WS, Noor AK (1995) Assessment of computational models for sandwich panels and shells. Comput Meth Appl Mech Eng 124:125–151
6. Donell LH (1976) Beams, plates and shells. McGraw-Hill, New York
7. Flügge W (1962) Statik und Dynamik der Schalen. Springer, Berlin
8. Flügge W (1967) Stresses in shells. Springer, Berlin
9. Gol'denveizer AL (1961) Theory of elastic shells (translated from the Russian by G. Herrmann). Pergamon Press, New York
10. Grigorenko YM, Gulyaev VI (1991) Nonlinear problems of shell theory and their solution methods (Review). Int Appl Mech 27(10):929–947
11. Korn GA, Korn TM (1961) Mathematical handbook for scientists and engineers. McGraw-Hill book company Inc., New York
12. Kraus H (1967) Thin elastic shells. Wiley, New York
13. Lamb H (1890) On the determination of an elastic shell. Proc London Math Soc 17:119–146
14. Leissa AW (1973) Vibration of shells. NASA SP-288, Washington, DC
15. Librescu L (1975) Elastostatics and kinetics of anisotropic and heterogeneous shell-type structures. Noordhoff Inter. Publishing
16. Lekhnitskii SG (1981) Theory of elasticity of an anisotropic body. English translation Mir Publishers
17. Love AEH. (1952) Mathematical theory of elasticity. Cambridge University Press, Cambridge
18. Mindlin RD (2006) An introduction to the mathematical theory of elastic plates. World Scientific Publishing Co. Pte. Ltd
19. Mushtari KhM, Galimov KZ (1957) Nonlinear theory of thin elastic shells. Russian translational series, NASA TT-F-62, Washington, DC, US Department of Commerce
20. Novozhilov VV (1964) Theory of thin elastic shells. P. Noordhoff, Groningen
21. Nowacki W (1963) Dynamics of elastic systems. Wiley, New York
22. Reddy JN (2003) Mechanics of laminated composite plates and shells: theory and analysis. CRC Press, Boca Raton
23. Reissner E, Arbor A, Edwards JW (1949) On the theory of thin elastic plates. Contributions to Appl Mech (H. Reissner Anniversary Volume) 231–247
24. Sanders LJr (1961) Nonlinear theories for thin shells. Harvard Press, Cambridge
25. Sokolnikoff IS (1956) Mathematical theory of elasticity. McGraw-Hill, New York
26. Timoshenko SP, Goodier JN (1970) Theory of elasticity. McGraw-Hill, New York
27. Timoshenko SP, Woinowsky-Krieger S (1959) Theory of plates and shells. McGraw-Hill, New York
28. Vlasov VZ (1964) General theory of shells and its application in engineering (English trans). NASA TT F-99, Washington, DC

Chapter 2
Discrete-Continuous Methods for Solution

Abstract Analytical-numerical methods for solving boundary-value and boundary-value eigenvalue problems for systems of ordinary differential equations and partial differential equations with variable coefficients are presented. In order to solve one-dimensional problems, the discrete-orthogonalization method is proposed. This approach is based on reducing the boundary-value problem to a number of Cauchy problems followed by their orthogonalization at some points of the integration interval which provides stability of calculations. In the case of the boundary-value eigenvalue problems, this approach is employed in combination with the incremental search method. In order to solve two-dimensional problems, the original system of partial differential system is reduced to systems of ordinary differential equations while making use of spline approximation and solving them by the discrete-orthogonalization method. Employing spline-functions is favorable, first, because of stability with respect to local disturbances, i.e., in contrast to polynomial approximation the spline behavior in the vicinity of a point does not influence the spline behavior as a whole; second, more satisfactory convergence is achieved, in contrast to the case of polynomials being applied as approximation functions; third, simplicity and convenience in calculation and implementation of spline-functions with the help of modern computers results. Besides, a nontraditional approach to solving problems of that class is proposed. It makes use of discrete Fourier series, i.e., Fourier series for functions specified on the discrete set of points. The two-dimensional boundary-value problem is solved by reducing it to a one-dimensional one after introducing auxiliary functions and separation of variables by using discrete Fourier series. Taking into account the calculation possibilities of modern computers, which make it possible to calculate a large number of series terms, the problem can be solved with high accuracy.

Keywords Discrete-orthogonalization method · Spline-collocation method · Discrete Fourier series method

A.Ya. Grigorenko et al., *Recent Developments in Anisotropic Heterogeneous Shell Theory*, SpringerBriefs in Continuum Mechanics,
DOI 10.1007/978-981-10-0353-0_2

2.1 Discrete-Orthogonalization Method

It should be noted that in studying mechanical behavior of the wide class of shell structures, a two-dimensional problem can be reduced to the systems of ordinary differential equations by separation of variables and by using Fourier series. However, during the numerical solution of one-dimensional boundary-value problems describing the stationary deformation of shells of revolution, boundary and local effects are immanent. These effects must be attributed to the rapid growth of the resolving functions and, as a result, gave rise to instability during the calculation process. From a mathematical point-of-view this means that, if the eigenvalues of the matrix of the system differ appreciably in the value of the real part, then when integrating, with an argument increasing as the result of loss of significant digits, the system of solution vectors for the Cauchy problems becomes almost linearly dependent. For this reason it is impossible to determine the constants of integration and the sought functions so that the boundary conditions on other end of the interval are satisfied with sufficient accuracy. It may occur that reliable digits will be absent in the solution at all. Attention to this phenomenon was paid in Collatz [3].

From the preceding, the necessity arises in constructing a stable numerical algorithm. To overcome these difficulties, the discrete-orthogonalization method is proposed in this book which, due to the orthogonalization of the vector-solutions of Cauchy problems at a finite number of points within the range of variation of the argument, makes it possible to obtain a stable calculation process. The principal idea of this approach was proposed by Bellman and Kalaba [2]. In the works of Grigorenko [6], Grigorenko and Vasilenko [11, 12], Grigorenko and Grigorenko [9] a computer-aided algorithm based on the above approach is proposed, which is effective and convenient in realization. The algorithm permits to solve the problems with high accuracy.

The idea of the discrete-orthogonalization method lies in the following. We consider the linear boundary-value problem

$$\frac{\mathrm{d}\bar{g}}{\mathrm{d}t} = A(t)\bar{g}(t) + \bar{f}(t), \quad a \le t \le b \tag{2.1.1}$$

with boundary conditions

$$B_1\bar{g}(a) = \bar{b}_1, \tag{2.1.2}$$

$$B_2\bar{g}(b) = \bar{b}_2, \tag{2.1.3}$$

where $\bar{g} = \{g_1, g_2, \ldots, g_n\}^{\mathrm{T}}$ is the column vector of the unknowns; $\bar{f}$ is the column vector of the right-hand part; $A(t)$ is the specified square matrix of the n th order; B_1 and B_2 are the specified rectangular matrixes of the $k \times n$th and $(n-k) \times n$th order, $(k < n)$; $\bar{b}_1$ and $\bar{b}_2$ are specified vectors.

We will seek for the solution of the boundary-value problem in the form

$$\bar{g}(t) = \sum_{j=1}^{m} C_j \bar{g}_j(t) + \bar{g}_{m+1}(t) \tag{2.1.4}$$

where $m = \min\{k, n-k\}$ (for determinacy it is assumed that $m = n - k$); $\bar{g}_j$ are the solutions of the Cauchy problems for the system (2.1.1) at $\bar{f} = 0$ with initial conditions satisfying the boundary conditions at the left end of the interval, $t = a$, for $\bar{b}_1 = 0$; $\bar{g}_{m+1}$ is the solution of the Cauchy problem for the system (2.1.1) with initial conditions satisfying the boundary conditions (2.1.2); m is the number of boundary conditions at the right end of the integration interval.

The above-mentioned requirements can be satisfied in the following manner. We present the boundary conditions at the point $-\cdots - w_{m+1,m-1}\bar{z}_{m-1} - w_{m+1,m}\bar{z}_m$ in expanded form:

$$\begin{aligned}
&b_{11}g_1 + b_{12}g_2 + \cdots + b_{1k}g_k + b_{1,k+1}g_{k+1} + \cdots + b_{1n}g_n = b_1,\\
&b_{21}g_1 + b_{22}g_2 + \cdots + b_{2k}g_k + b_{2,k+1}g_{k+1} + \cdots + b_{2n}g_n = b_2,\\
&\cdots\cdots\cdots\cdots\cdots\cdots\cdots\cdots\cdots\cdots\cdots\cdots\\
&\bar{z}_{m+1} = \bar{u}_{m+1} - w_{m+1,1}\bar{z}_1 - w_{m+1,2}\bar{z}_2.
\end{aligned} \tag{2.1.5}$$

Now we suppose that the coefficients of the first k columns in Eq. (2.1.5) form a nonsingular matrix and transfer the remaining columns to the right-hand side. Then the conditions (2.1.5) become:

$$\begin{aligned}
&b_{11}g_1 + b_{12}g_2 + \cdots + b_{1k}g_k = b_1 - b_{1,k+1}g_{k+1} - \cdots - b_{1n}g_n,\\
&b_{21}g_1 + b_{22}g_2 + \cdots + b_{2k}g_k = b_2 - b_{2,k+1}g_{k+1} - \cdots - b_{2n}g_n,\\
&\cdots\cdots\cdots\cdots\cdots\cdots\cdots\cdots\cdots\cdots\cdots\cdots\\
&b_{k1}g_1 + b_{k2}g_2 + \cdots + b_{kk}g_k = b_k - b_{k,k+1}g_{k+1} - \cdots - b_{kn}g_n.
\end{aligned} \tag{2.1.6}$$

By assigning successively values of columns of the unit matrix to the components $g_{k+1}, g_{k+2}, \ldots, g_n$ and by assuming that $-\cdots - w_{m+1,m-1}\bar{z}_{m-1} - w_{m+1,m}\bar{z}_m$, we define initial conditions for $\bar{g}_j (j = 1, 2, \ldots, m)$; if $g_{k+1} = g_{k+2} = \cdots = g_n = 0$, then we find initial conditions for $\bar{g}_{m+1}$.

The discrete-orthogonalization method makes it possible to provide a stable computational process by the way of orthogonalization of the solution vectors of the Cauchy problem at a finite number of points of the interval where the argument varies. Let us divide the segment $[a, b]$ into parts by integration points $t_s (s = 0, 1, \ldots, N)$ in such a way that $t_0 = a, t_N = b$. From these points let us choose orthogonalization points $T_i (i = 0, 1, \ldots, M)$. The selection of the indicated points is dictated usually by the level of desired accuracy of the problem being solved as for the rest it is arbitrary.

Let the solutions of the Cauchy problem, which we designate as $\bar{u}_r(T_i)(r = 1, 2, \ldots, m + 1)$, be found at the point T_i by the some numerical method, for example, the Runge-Kutta method. Then, before orthogonalization, we have the following vectors at the point T_i:

$$\bar{u}_1(T_i), \bar{u}_2(T_i), \ldots, \bar{u}_m(T_i), \bar{u}_{m+1}(T_i). \tag{2.1.7}$$

Let us orthonormalize the vectors $\bar{u}_j(T_j)$ $(j = 1, 2, \ldots, m)$ at the point T_i and designate them by

$$\bar{z}_1(T_i), \bar{z}_2(T_i), \ldots, \bar{z}_m(T_i). \tag{2.1.8}$$

The vectors $\bar{z}_i$ can be expressed in terms of the vectors $\bar{u}_i$ as follows:

$$\bar{z}_r = \frac{1}{w_{rr}}\left(\bar{u}_r - \sum_{j=1}^{r-1} w_{rj}\,\bar{z}_j\right) \quad (r = 1, 2, \ldots, m), \tag{2.1.9}$$

where

$$w_{rj} = (\bar{u}_r, \bar{z}_j), j < r, w_{rr} = \sqrt{(\bar{u}_r, \bar{z}_r) - \sum_{j=1}^{r-1} w_{rj}^2}.$$

The vector $\bar{z}_{m+1}$ is not subjected to normalization and can be calculated by the formula

$$\bar{z}_{m+1} = \bar{u}_{m+1} - \sum_{j=1}^{m} w_{m+1,j}\,\bar{z}_j. \tag{2.1.10}$$

By using (2.1.9) and (2.1.10), we have for $t = T_i$:

$$\begin{aligned}
w_{11}\bar{z}_1 &= \bar{u}_1, \\
w_{22}\bar{z}_2 &= \bar{u}_2 - w_{21}\bar{z}_1, \\
w_{33}\bar{z}_3 &= \bar{u}_3 - w_{31}\bar{z}_1 - w_{32}\bar{z}_2, \\
&\cdots\cdots\cdots\cdots\cdots\cdots \\
w_{mm}\bar{z}_m &= \bar{u}_m - w_{m1}\bar{z}_1 - w_{m2}\bar{z}_2 - \cdots - w_{m,m-1}\bar{z}_{m-1}, \\
\bar{z}_{m+1} &= \bar{u}_{m+1} - w_{m+1,1}\bar{z}_1 - w_{m+1,2}\bar{z}_2 - \cdots - w_{m+1,m-1}\bar{z}_{m-1} - w_{m+1,m}\bar{z}_m.
\end{aligned} \tag{2.1.11}$$

After transformations, we obtain a matrix equation:

$$\begin{vmatrix} \bar{u}_1(T_i) \\ \bar{u}_2(T_i) \\ \cdot \\ \cdot \\ \cdot \\ \bar{u}_m(T_i) \\ \bar{u}_{m+1}(T_i) \end{vmatrix} = \Omega_i \begin{vmatrix} \bar{z}_1(T_i) \\ \bar{z}_2(T_i) \\ \cdot \\ \cdot \\ \cdot \\ \bar{z}_m(T_i) \\ \bar{z}_{m+1}(T_i) \end{vmatrix}, \tag{2.1.12}$$

where:

$$\Omega_i = \Omega(T_i) = \begin{vmatrix} w_{11}(T_i) & 0 & 0 & \cdot\;\cdot\;\cdot & 0 \\ w_{21}(T_i) & w_{22}(T_i) & 0 & \cdot\;\cdot\;\cdot & 0 \\ w_{31}(T_i) & w_{32}(T_i) & w_{33}(T_i) & \cdot\;\cdot\;\cdot & 0 \\ \cdots & \cdots & \cdots & \cdot\;\cdot\;\cdot & \cdot \\ w_{m1}(T_i) & w_{m2}(T_i) & w_{m3}(T_i) & \cdot\;\cdot\;\cdot & 0 \\ w_{m+1,1}(T_i) & w_{m+1,2}(T_i) & w_{m+1,3}(T_i) & \cdot\;\cdot\;\cdot & 1 \end{vmatrix}. \tag{2.1.13}$$

The vectors $\bar{z}_r(T_i)$ are the initial values of the Cauchy problems for homogeneous $(r = 1, 2, \ldots, m)$ and inhomogeneous $(r = m+1)$ systems of differential equations (2.1.1) on the interval $T_i \le t \le T_{i+1}$. The solution of this system, which satisfies boundary conditions at the left end of the interval according to Eq. (2.1.2) at each orthogonalization point T_i, can be written as follows:

before orthogonalization

$$\bar{g}(T_i) = \sum_{j=1}^{m} C_j^{(i-1)}\, \bar{u}_j(T_i) + \bar{u}_{m+1}(T_i), \tag{2.1.14}$$

and after orthogonalization:

$$\bar{g}(T_i) = \sum_{j=1}^{m} C_j^{(i)} \bar{z}_j(T_i) + \bar{z}_{m+1}(T_i). \tag{2.1.15}$$

The solution of the system (2.1.7) in the interval $T_i \le t \le T_{i+1}$ can be written as

$$\bar{g}(t) = \sum_{j=1}^{m} C_j^{(i)}\, \bar{z}_j(t) + \bar{z}_{m+1}(t). \tag{2.1.16}$$

Having integrated the Cauchy equations for the last section $T_{M-1} \leq t \leq T_M$ and after having orthogonalizing them at the point T_M by (2.1.15), we get:

$$\bar{g}(T_M) = \sum_{j=1}^{m} C_j^{(M)} \bar{z}_j(T_M) + \bar{z}_{m+1}(T_M). \tag{2.1.17}$$

By satisfying the boundary conditions at the right end of the integration interval, we get a system of m linear algebraic equations for determining the unknowns $C_j^{(M)} (j = 1, 2, \ldots, m)$. After finding $C_j^{(M)}$, the solution of the boundary-value problem (2.1.11)–(2.1.13) at the point $t = T_M$ becomes (2.1.17). At this stage the direct procedure of the problem solution finishes.

In the case of reverse procedure, we determine the constants $C_j^{(i)}$ $(j = 1, 2, \ldots, m)$ from the values $C_j^{(i-1)}$ beginning with $i = M$. In order to do this, we equate the right-hand sides of the expressions (2.1.14) and (2.1.16):

$$\sum_{j=1}^{m} C_j^{(i-1)} \bar{u}_j(T_i) + \bar{u}_{m+1}(T_i) = \sum_{j=1}^{m} C_j^{(i-1)} \bar{z}_j(T_i) + \bar{z}_{m+1}(T_i). \tag{2.1.18}$$

By substituting their values from (2.1.12) instead of $\bar{u}_j$, for $t = T_i$, we have:

$$\begin{aligned} & C_1^{(i-1)} w_{11}\bar{z}_1 + C_2^{(i-1)}(w_{21}\bar{z}_1 + w_{22}\bar{z}_2) \\ & \quad + C_3^{(i-1)}(w_{31}\bar{z}_1 + w_{32}\bar{z}_2 + w_{33}\bar{z}_3) + \cdots \\ & \quad + C_m^{(i-1)}(w_{m1}\bar{z}_1 + w_{m2}\bar{z}_2 + \cdots + w_{mm}\bar{z}_m) \\ & \quad + (w_{m+1,1}\bar{z}_1 + w_{m+1,2}\bar{z}_2 + \cdots + w_{m+1,m}\bar{z}_m + \bar{z}_{m+1}) \\ & \quad = C_1^{(i)}\bar{z}_1 + C_2^{(i)}\bar{z}_2 + \cdots + C_m^{(i)}\bar{z}_m + \bar{z}_{m+1}. \end{aligned} \tag{2.1.19}$$

By equating the coefficients of the vectors $\bar{z}_j$ $(j = 1, 2, \ldots, m+1)$ in Eq. (2.1.19) we find:

$$\Omega_i' \bar{C}^{(i-1)} = \bar{C}^{(i)} \quad (i = 1, 2, \ldots, M), \tag{2.1.20}$$

or

$$\bar{C}^{(i-1)} = \left[\Omega_i'\right]^{-1} \bar{C}^{(i)},$$

where Ω_i' is the transposed matrix to (2.1.19) and $\bar{C}^{(i)}$ is the column vector with components $(j = 1, 2, \ldots, m)$.

Hence, by using (2.1.20), we find the constants $C_j^{(i)}$ at all the points, beginning with $i = M$. The solutions $\bar{g}(T_i)$ of the boundary-value problem are found with Eq. (2.1.16). If the presented algorithm is implemented on a computer, it is necessary to keep the information on the matrices Ω_i and vectors $\bar{z}_r, (r = 1, 2, \ldots, m+1)$.

For practical problem solving, the information obtained at all points of the orthogonalization is usually not used with the exception of the values of the sought functions at the so-called points of generation of results. Usually the number of these points is far less than the number of orthogonalization points, which makes it possible to cut down the volume of kept information considerably by using the following technique.

Let T_{i-1} and T_{i+p} be the points of generation of results. From Eq. (2.1.20) we obtain:

$$\Omega'_{i+p}\Omega'_{i+p-1}\cdots\Omega'_i \quad \bar{C}^{(i-1)} = \bar{C}^{(i+p)} \tag{2.1.21}$$

or:

$$\left(\prod_{j=0}^{p}\Omega_{i+j}\right)' \quad \bar{C}^{(i-1)} = \bar{C}^{(i+p)},$$

where we find:

$$\bar{C}^{(i-1)} = \left[\left(\prod_{j=0}^{p}\Omega_{i+j}\right)'\right]^{-1} \bar{C}^{(i+p)}. \tag{2.1.22}$$

Thus in order to define the vectors $\bar{C}^{(i-1)}$ it is necessary to keep the information on the product of matrices $\prod_{j=0}^{p}\Omega_{i+j}$. This provides a means for considerable savings of computer memory.

By increasing the number of orthogonalization points, we can essentially decrease the calculation error caused by the rigidity of the equation set. Besides, the boundary-value problem can be solved by "from the left to the right" or "from the right to the left" methods.

In the case of a boundary-value eigenvalue problem for the systems of differential equations:

$$\frac{\mathrm{d}\bar{Y}}{\mathrm{d}x} = A(x,\omega)\bar{Y}(x), \quad (0 \le x \le a). \tag{2.1.23}$$

with the boundary conditions:

$$B_1\bar{Y}(0) = \bar{0}, \tag{2.1.24}$$

$$B_2\bar{Y}(a) = \bar{0}, \tag{2.1.25}$$

where $\bar{Y} = \{y_1, y_2, \ldots, y_n\}^T$ is the vector-column, $A(x, \omega)$ is a quadratic matrix of nth order, B_1 and B_2 are matrices of the order $k \times n$ and $(n - k) \times n$ $(k < n)$, respectively. The discrete-orthogonalization method in combination with the incremental search method is applied to determine the eigenvalues and the eigenvectors for the problem defined by Eqs. (2.1.23) and (2.1.24). All features of the approach are outlined in detail in Grigorenko [4], Grigorenko and Efimova [5], and Gigorenko et al. [10].

2.2 Spline-Collocation Method

Currently for solving problems of computational mathematics, mathematical physics, and mechanics spline-functions are widely used. This is due to various advantages of the spline-approximation technique when compared to others. As basic advantages, the following ones should be mentioned: First, the stability of splines with respect to local disturbances, i.e., the behavior of the spline near a point does not affect the behavior of the spline as a whole as, for instance, in the case of polynomial approximation; second, fast convergence of the spline-interpolation in contrast to the polynomial one; third, simplicity and convenience in realization of algorithms for constructing and calculating splines on personal computers. The use of spline functions in various variational, projective, and other discrete-continuous methods makes it possible to obtain appreciable results in comparison with those obtained with the classical apparatus of polynomials, to simplify essentially their numerical realization, and to obtain the desired solution with a high degree accuracy (Grigorenko and Berenov [8], Grigorenko and Efimova [5]).

2.2.1 Basic Information on Spline-Functions and Their Application for Solving Boundary-Value Problems

A spline is a function "sewn' of the "fragments" of generalized polynomials with respect to a specified basis.

Among polynomial splines the most widely used are those whose base functions consist of $1, x, x^2, \ldots$. Later on we will restrict ourselves to polynomial splines (Alberg et al. [1]).

Separated by the mesh distance Δ we discretize the segment $[a, b]$ such that $a = x_0 < x_1 < x_2 \ldots < x_N = b$. We denote by P_m the set of polynomials with a power not more than m and by $C^{(k)}[a, b]$ the set of functions, which are k-times continuously differentiable in $[a, b]$. The function $S_{m,k}(x)$ is the spline function of m-th order and k, $1 \le k \le m$ is the difference between m and the order of the highest continuous derivative with nodes on the mesh Δ, if:

(a) $S_{m,k}(x) \in P_m$ at $x \in [x_i, x_{i+1}]$, $\quad i = 0, 1, 2, \ldots, N-1$;
(b) $S_{m,k}(x) \in C^{(m-k)}[a, b]$.

From the above it follows that the spline $S_{m,k}(x)$ has continuous derivatives up to the $(m-k)$th order and the $(m-k+1)$th derivative can have a discontinuity in $[a, b]$.

We designate the set of splines corresponding to the above by $S_{m,k}(\Delta)$. The simplest example of a spline is the Heaviside function:

$$\theta(x) = \begin{cases} 1 & x \ge 0, \\ 0 & x < 0. \end{cases}$$

Another example is a truncated power function:

$$x_+^m = x^m \theta(x) = \begin{cases} x^m & x \ge 0, \\ 0 & x < 0. \end{cases}$$

The functions $\theta(x)$ and x_+^m are splines of zero power and of mth power with one node $x = 0$, respectively.

If the value $u (a < u < b)$ is fixed, then the function $(x-u)_+^m$ belongs to $C^{(m-1)}[a, b]$ and can be considered as the mth power spline with the node $x = u$. Therefore the truncated power functions $(x - x_i)_+^{\alpha}$ related to the points of the mesh Δ, at $m - k + 1 \le \alpha \le m$ belong to the set $S_{m,k}(\Delta)$.

The functions $B_m^i(x) = \frac{x_{i+m+1} - x_i}{m+1} \tilde{B}_m^i(x)$ and $(x - x_i)_+^{\alpha'}$ $(\alpha' = n - k + 1, \ldots, N;$ $1 \le k \le N + 1;$ $i = 1, 2, \ldots, N - 1)$ are linearly independent and form a base in the space $S_{m,k}(\Delta)$ of dimension $(m+1) + k(N-1)$.

Depending on the choice of base functions, we will have different analytical representation of splines. By using the truncated power functions, it is possible to represent the mth-power spline of the defect k with nodes on the mesh Δ unambiguously as:

$$S_{m,k}(x) = \sum_{v=0}^{m} c_v (x-a)^v + \sum_{i=1}^{N-1} \sum_{j=0}^{k-1} \alpha_{ij} (x - x_i)_+^{m-j}, \tag{2.2.1}$$

where:

$$C_v = S_{m,k}(a)/v!, \quad v = 0, 1, 2, \ldots, m,$$
$$\alpha_{ij} = \left[S_{m,k}^{(m-j)}(x_j + 0) - S_{m,k}^{(m-j)}(x_j - 0) \right] / (m-j)!,$$
$$i = 1, 2, \ldots, N-1, \quad j = 0, 1, \ldots, k-1.,$$

Later on we will consider only splines at $k = 1$, denoting them in terms of $S_m(x)$.

Let a new mesh be defined by $\bar{\Delta} : a = \bar{x}_0 < \bar{x}_1 < \bar{x}_2 < \cdots < \bar{x}_N = b$ and real numbers denoted by $y_i (i = \overline{0, N})$. The spline $S_m(x)$ is called interpolation polynomial spline on the mesh $\bar{\Delta}$ for interpolation of the function $f(x)$, if:

(a) $S_m(x) \in P_m$ at $x \in [\bar{x}_i, \bar{x}_{i+1}], \quad i = 0, 1, 2, \ldots, N - 1;$
(b) $S_m(x) \in C^{(m-1)}[a, b];$
(c) $S_m(\bar{x}_i) = y_i = f(\bar{x}_i), \quad i = 0, 1, \ldots, N$

In this case, the nodes of the mesh Δ are called spline nodes and the nodes of the mesh $\bar{\Delta}$ are referred to as interpolation nodes. We suppose that for splines of uneven power the meshes Δ and $\bar{\Delta}$ coincide.

The spline $S_m(x)$ from the set $S_m(\Delta)$ is "glued" to $N - 1$ nodes $\bar{x}_1, \bar{x}_2, \ldots, \bar{x}_{N-1}$ to the $(m - 1)$ th derivative of the N algebraic polynomials of power m inclusively. For this reason it has $N(m + 1) - (N - 1)m = N + m$ free parameters. These parameters should be used in such a way as to satisfy interpolation equations of the $S^{(v)}(\tau) = y_\tau^{(v)}$ type, where for $a < \tau < b$, we adopt $0 \leq v \leq m - 1$. At $\tau = a$ and $\tau = b$ these conditions are related to boundary conditions assuming that $0 \leq v \leq m$. The boundary conditions can be specified also in the form of linear equations, which relate values $S_m^{(v)}(a)$ and $S_m^{(\mu)}(b)(0 \leq v,\ \mu \leq m)$.

The problem regarding the existence of the interpolation spline $S_m(x)$ satisfying interpolation and boundary conditions can be correctly posed if the number of equations by which these conditions are specified, does not exceed the number of free parameters, i.e., $N + m$. If the spline $S_m(x)$ interpolating the function $f(x)$ under certain boundary conditions exists, it can be expressed in the form of linear combinations of base functions, for example, in the form of Eq. (2.2.1). Coefficients appearing in this expression are determined from the set of equations that can be obtained by substituting the expression into interpolation and boundary conditions. The corresponding solution is of rather complicated nature despite the fact that the base functions are simple. For this reason, other base functions are used in practice, which make it possible to represent the interpolation spline in such a form, that the coefficients are simply expressed in terms of the interpolation function or its derivatives.

The coefficients have the simplest form if the spline is represented by fundamental splines, i.e., in a form similar to the known Lagrange or Hermite formulae. In this case the coefficients are equal to values of the interpolated function and its derivatives. Let interpolation conditions be specified in the points $a = x_0 < x_1 < x_2 \ldots < x_N = b$. The value of the interpolation spline is one if the interpolation condition is satisfied, otherwise its value is zero. Then under the specified conditions we have:

$$S(x) = \sum_{i=0}^{N} f(x_i) F_m^i(x) + \sum_{\mu \in I_a} f^{(\mu)}(a) F_m^a(x) + \sum_{v \in I_b} f^{(v)}(b) F_m^b(x). \tag{2.2.2}$$

In this case, the fundamental splines have the form $F_m^i(x_p) = \delta_{ip}$ for $x_p \in \Delta$, if $i = 0, 1, 2, \ldots, N$, where δ_{ip} is the Kronecker delta, i.e.:

$$\delta_{ip} = \begin{cases} 1 & i = p, \\ 0 & i \neq p. \end{cases}$$

The fundamental splines $F_m^a(x)$ and $F_m^b(x)$ correspond to specified boundary conditions. The sets of integer numbers I_a and I_b define the number of boundary conditions at points $x = a$ and $x = b$, respectively.

The representation of the spline in the form of Eq. (2.2.2) is known as Lagrange's interpolation formula for splines. At $N - m = 0$ $F_m^i(x)$ are the fundamental polynomials, and $S(x)$ is the Lagrange interpolation polynomial.

2.2.2 *B-Splines*

Despite of the sufficiently explicit form of representation of the interpolation spline (2.2.2) (from the standpoint of coefficients), computational difficulties arise in its construction. In order to avoid this, B-splines are used. We augment the mesh $\Delta : a = x_0 < x_i < \cdots < x_N = b$ by auxiliary points $x_{-m} < \cdots < x_{-1} < a$, $b < x_{N+1} < \cdots < x_{N+m}$ and consider the mesh $\Delta_1 : x_{-m} < \cdots < x_{-1} < x_0 < x_1 < \cdots < x_{N-1} < x_N < x_{N+1} < \cdots < x_{N+m}$.

Let us consider the functions $\phi_m(x, t) = (-1)^{m+1}(m+1)(x-t)_+^m$ and construct the divided difference of the $(m+1)$ th order for values of the argument $t = x_i, \ldots, x_{i+m+1}$. As a result, we obtain the following functions of the variable x:

$$\tilde{B}_m^i = \phi_m[x; x_i, \ldots, x_{i+m+1}], \quad i = -m, \ldots, N-1. \tag{2.2.3}$$

These functions are called base splines or B-splines of the mth power and are splines of the mth power of the $k = 1$ on the broadened mesh Δ_1. By using the identity $(x-t)_+^m = (x-t)^m + (-1)^{m+1}(t-x)_+^m$ we can obtain another form of the expression (2.2.3):

$$\tilde{B}_m^i(x) = (m+1) \sum_{p=i}^{i+m+1} \frac{(x_p - x)_+^m}{\omega'_{m+1,i}(x_p)} \quad (i = -m, \ldots, N-1), \tag{2.2.4}$$

where $\omega_{m+1,i}(t) = \prod_{j=i}^{i+m+1} (t - x_j)$.

In practical calculations, it is convenient to use normalized B-splines in the form

$$B_m^i(x) = \frac{x_{i+m+1} - x_i}{m+1} \tilde{B}_m^i(x) \tag{2.2.5}$$

instead of the B-splines. For the normalized B-splines, the recurrence formula

$$B_m^i(x) = \frac{x - x_i}{x_{i+m} - x_i} B_{m-1}^i(x) + \frac{x_{i+m+1} - x}{x_{i+m+1} - x_{i+1}} B_{m-1}^{i+1}(x) \tag{2.2.6}$$

holds, which can be used as a definition for the B-splines. Then:

$$B_0^i(x) = \begin{cases} 1 & x \in [x_i, x_{i+1}), \\ 0 & x \notin [x_i, x_{i+1}). \end{cases}$$

The functions $B_m^i(x)$ are splines of the mth power of $k = 1$. Besides, the system of functions $B_m^i(x)$ $i = -m, \ldots, N-1$ is linearly independent and forms a base in the space of splines $S_m(\Delta)$. It indicates that each spline $S_m(x) \in S_m(\Delta)$ can be uniquely presented by

$$S_m(x) = \sum_{i=-m}^{N-1} b_i B_m^i(x), \tag{2.2.7}$$

where b_i are some constants.

The splines $B_m^i(x)$ have the following properties:

(a) $B_m^i(x) > 0$ for $x \in [x_i, x_{i+1})$, $B_m^i(x) \equiv 0$ for $x \notin [x_i, x_{i+1})$;

(b) $\int\limits_{-\infty}^{\infty} B_m^i(x)dx = \frac{x_{i+m+1} - x_i}{m+1}$.

Let us consider the uniformly widened mesh $\Delta' : x_{-m} < \cdots < x_{-1} < x_0 < \cdots < x_N < x_{N+1} < \cdots < x_{N+m}$ $(x_{k+1} - x_k = h = \text{const})$. Let us construct three odd power B-splines. In this case, we will numerate them by the middle node of carriers. For this reason we will denote the odd-power B-splines by $B_m^i(x)$ instead of $B_m^{i-(m+1)/2}(x)$, i.e., the spline numeration shifts to the right by the amount $(m+1)/2$.

Thus, we have first-power B-splines:

$$B_1^i(x) = \begin{cases} 0, & -\infty < x < x_{i-1}, \\ t, & x_{i-1} \le x \le x_i, \\ 1 - t, & x_i \le x < x_{i+1}, \\ 0, & x_{i+1} \le x < \infty; \end{cases} \tag{2.2.8}$$

third-power B-splines:

$$B_3^i(x) = \frac{1}{6}\begin{cases} 0, & -\infty < x < x_{i-2}, \\ t^3, & x_{i-2} \le x < x_{i-1}, \\ -3t^3 + 3t^2 + 3t + 1, & x_{i-1} \le x < x_i, \\ 3t^3 - 6t^2 + 4, & x_i \le x < x_{i+1}, \\ (1-t)^3, & x_{i+1} \le x < x_{i+2}, \\ 0 & x_{i+2} \le x < \infty; \end{cases} \tag{2.2.9}$$

fifth-power B-splines:

$$B_5^i(x) = \frac{1}{120}\begin{cases} 0, & -\infty < x < x_{i-3}, \\ t^5, & x_{i-3} \le x < x_{i-2}, \\ -5t^5 + 5t^4 + 10t^3 + 10t^2 + 5t + 1, & x_{i-2} \le x < x_{i-1}, \\ 10t^5 - 20t^4 - 20t^3 + 20t^2 + 50t + 26, & x_{i-1} \le x < x_i, \\ -10t^5 + 30t^4 - 60t^2 + 66, & x_i \le x < x_{i+1}, \\ 5t^5 - 20t^4 + 20t^3 + 20t^2 - 50t + 26 & x_{i+1} \le x < x_{i+2} \\ (1-t)^5, & x_{i+2} \le x < x_{i+3}, \\ 0, & x_{i+3} \le x < \infty, \end{cases} \tag{2.2.10}$$

where $t = \frac{x - x_k}{h}$ on the interval $[x_k, x_{k+1}]$, $k = i = \overline{-\frac{m+1}{2}, \ i + \frac{m+1}{2} - 1}$; $i = \overline{-\frac{m+1}{2} + 1, \ N + \frac{m+1}{2} - 1}; m = 1, 3, 5$.

Tables 2.1 and 2.2 summarize values of the splines $B_3^i(x)$ and $B_5^i(x)$ and their derivatives at the nodes of splines.

Table 2.1 Summarize value of the spline $B_3^i(x)$ and its derivatives

x	B_3^i	$(B_3^i)'$	$(B_3^i)''$
x_{i-2}	0	0	0
x_{i-1}	1/6	$1/(2h)$	$1/(2h)$
x_i	4/6	0	$-2/h^2$
x_{i+1}	1/6	$-1/(2h)$	$1/h^2$
x_{i+2}	0	0	0

Table 2.2 Summarize value of the spline $B_5^i(x)$ and its derivatives

x	B_5^i	$(B_5^i)'$	$(B_5^i)''$	$(B_5^i)^{III}$	$(B_5^i)^{IV}$
x_{i-3}	0	0	0	0	0
x_{i-2}	$\frac{1}{120}$	$\frac{1}{24h}$	$\frac{1}{6h^2}$	$\frac{1}{2h^3}$	$\frac{1}{h^4}$
x_{i-1}	$\frac{26}{120}$	$\frac{10}{24h}$	$\frac{2}{6h^2}$	$-\frac{1}{h^3}$	$-\frac{4}{h^4}$
x_i	$\frac{66}{120}$	0	$-\frac{1}{h^2}$	0	$\frac{6}{h^4}$
x_{i+1}	$\frac{26}{120}$	$-\frac{10}{24h}$	$\frac{2}{6h^2}$	$\frac{1}{h^3}$	$-\frac{4}{h^4}$
x_{i+2}	$\frac{1}{120}$	$-\frac{1}{24h}$	$\frac{1}{6h^2}$	$-\frac{1}{2h^3}$	$\frac{1}{h^4}$
x_{i+3}	0	0	0	0	0

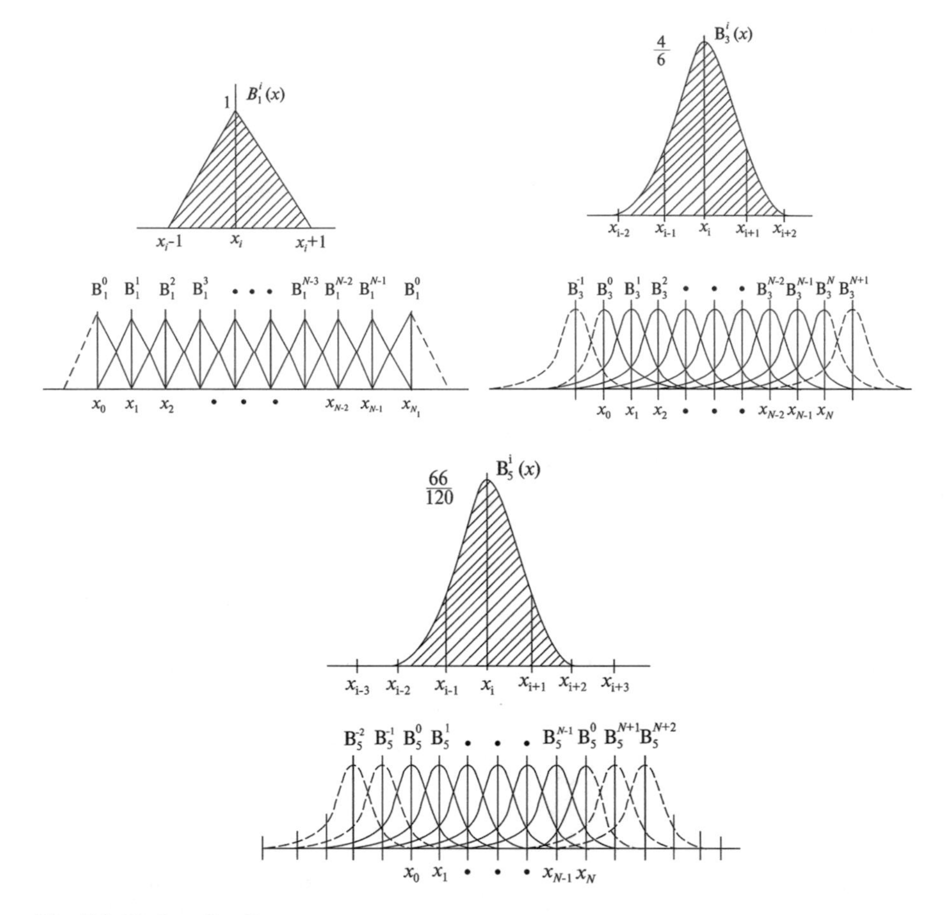

Fig. 2.1 Various B-splines

Figure 2.1 shows B-splines as well as the related bases of B-splines in the spaces $S_m(\Delta)$ $(m = 1,\ 3,\ 5)$.

2.2.3 *The Spline-Collocation Method*

We now turn to the following boundary-value problem:

$$Ly \equiv y'' + p(x)y' + q(x)y = f(x), \quad x \in [a, b], \tag{2.2.11}$$

$$\alpha_1 y(a) + \beta_1 y'(a) = \gamma_1, \alpha_2 y(b) + \beta_2 y'(b) = \gamma_2. \tag{2.2.12}$$

For its solution we will specify the mesh $\Delta : a = x_0 < x_1 < \cdots < x_N = b$ on the segment [*a*, *b*]. We look for an approximate solution of the boundary-value problem (2.2.11), (2.2.12) in the form of a cubic spline *S*(*x*) of the class C^2 with nodes on the mesh Δ. Let us choose independently of the method the points $\xi_k \in [a, b]$ ($k = 0, 1, 2, \ldots, N$) on the segment [*a*, *b*]. These points are the collocation points. By substituting the spline $S(x)$ into Eqs. (2.2.11) and (2.2.12), we will require that the residual of Eq. (2.2.11) is equal to zero at the collocation points. Then:

$$S''(\xi_k) + p(\xi_k)S'(\xi_k) + q(\xi_k)S(\xi_k) = f(\xi_k), \quad k = 0, 1, 2, \ldots, N \tag{2.2.13}$$

$$\alpha_1 S(a) + \beta_1 S'(a) = \gamma_1, \tag{2.2.14}$$

$$\alpha_2 S(b) + \beta_2 S'(b) = \gamma_2. \tag{2.2.15}$$

This system has $N + 3$ unknown coefficients in case of a cubic spline $S(x)$ of the class C^2. Since the chosen spline $S(x)$ satisfies two boundary conditions at the points $x = a$ and $x = b$, the number of collocation nodes should be equal to $N + 1$. Then the system of algebraic equations (2.2.13)–(2.2.15) will include $N + 3$ equations and the same number of unknowns. Having solved this system and determined the sought coefficients, we obtain the analytical solution of the boundary-value problem (2.2.11) and (2.2.12). This method is known as spline-collocation method.

The collocation points are situated on the segment $[a, b]$ in a prescribed manner. Since the solution of the problem is sought in the form of cubic splines, the segment $[x_i, x_{i+1}]$ cannot include more than three nodes. The structure of the system (2.2.13) and (2.2.14) depends on the choice of collocation points. In solving the boundary-value problem (2.2.12) and (2.2.13), it is expedient to use B-splines. To do this, we broaden the uniform mesh Δ on the segment $[a, b]$ by using the nodes $x_{-3} < x_{-2} < x_{-1} < x_0$, $x_N < x_{N+1} < x_{N+2} < x_{N+3}$. Then the third-power system of B-splines $B_3^i(x), i = -1, 0, \ldots, N + 1$ on the broadened mesh Δ_1 forms a basis in the space $S_3(\Delta)$. We will search for solutions of the problem (2.2.11) and (2.2.18) in the form

$$S(x) = \sum_{i=-1}^{N+1} b_i B_3^i(x). \tag{2.2.16}$$

We consider the case when the collocation nodes coincide with the spline nodes of the mesh Δ, i.e., $\xi_i = x_i (i = 0, 1, \ldots, N)$. Having substituted (2.2.16) into (2.2.11), and after considering the spline properties and their values at nodal points (see Table 2.2), we have:

$$b_{i-1}A_i + b_i C_i + b_{i+1}B_i = D_i, \quad i = 0, 1, \ldots, N, \tag{2.2.17}$$

where:

$$A_i = \frac{1}{3h}\left(1 - \frac{1}{2}p(x_i)h + \frac{1}{6}q(x_i)h^2\right),$$
$$B_i = \frac{1}{3h}\left(1 + \frac{1}{2}p(x_i)h + \frac{1}{6}q(x_i)h^2\right),$$
$$C_i = -A_i - B_i + \frac{1}{3}q(x_i)h, \quad D_i = \frac{1}{3}f(x_i)h.$$

The boundary conditions (2.2.12) of the boundary problem while allowing for (2.2.16) are rearranged to yield the equations:

$$\begin{aligned} &b_{-1}A_{-1} + b_0C_{-1} + b_1B_{-1} = D_{-1},\\ &b_{N-1}A_{N+1} + b_NC_{N+1} + b_{N+1}B_{N+1} = D_{N+1} \end{aligned} \tag{2.2.18}$$

where:

$$\begin{aligned} &A_{-1} = \alpha_1 h - 3\beta_1, && C_{-1} = 4\alpha_1 h, && B_{-1} = \alpha_1 h + 3\beta_1,\\ &A_{n+1} = \alpha_2 h - 3\beta_2, && C_{N+1} = 4\alpha_2 h, && B_{N+1} = \alpha_2 h + 3\beta_2,\\ &D_{-1} = 6\gamma_1 h, && D_{N=1} = 6\gamma_2 h. \end{aligned}$$

Equations (2.2.17) and (2.2.18) form a system of $N+3$ algebraic equations with respect to the unknowns b_i. By excluding the unknowns b_{-1} and b_{N+1} from the system (2.2.17) with the help of Eq. (2.2.18) we obtain the following system of equations with a tridiagonal matrix:

$$\begin{aligned} &b_0\tilde{C}_0 + b_1\tilde{B}_0 = \tilde{D}_0,\\ &b_{i-1}A_{i-1} + b_iC_i + b_{i+1}B_i = D_i \quad (i = 1, 2, \ldots, N-1),\\ &b_{N-1}\tilde{A}_N + b_N\tilde{C}_N = \tilde{D}_N, \end{aligned} \tag{2.2.19}$$

where:

$$\tilde{C}_0 = C_0 - \frac{C_{-1}A_0}{A_{-1}}, \ \tilde{B}_0 = \tilde{B}_0 - \frac{B_{-1}A_0}{A_{-1}}, \ \tilde{D}_0 = D_0 - \frac{D_{-1}A_0}{A_{-1}},$$
$$\tilde{A}_N = A_N - \frac{A_{N+1}B_N}{B_{N+1}}, \ \tilde{C}_N = C_N - \frac{C_{N+1}B_N}{B_{N+1}}, \ \tilde{D}_N = D_N - \frac{D_{N+1}B_N}{B_{N+1}}.$$

If the conditions

$$\beta_1 \le 0, \quad \beta_2 \ge 0, \quad \alpha_j \ge 0, \quad |\alpha_j| + |\beta_i| \ne 0, \quad j = 1, 2, \quad f(x) \le q < 0$$

are met and if the step h is reasonably small, such that

$$1-\frac{1}{2}p(x_i)h+\frac{1}{6}q(x_i)h^2\geq 0,\quad 1+\frac{1}{2}p(x_i)h+\frac{1}{2}q(x_i)h^2\geq 0,$$

then the system (2.2.19) would be a system with diagonal predominance.

If the collocation nodes do not coincide with the nodes of the partition of the segment $[a,b]$, then the structure of the system of algebraic equations will be determined by the arrangement of the collocation nodes. If each segment $[x_i,x_{i+1}]$ has not more than two collocation nodes, then the matrix of the system of equations will show a five-band structure.

Let us consider the case, when the number of nodes of the mesh $\varDelta$ is even, i.e., $N=2n+1$ and collocation nodes $\ldots\xi_k,\ k=0,1,\ldots,N$ satisfy the conditions

$$\xi_{2i}\in[x_{2i},x_{2i+1}],\quad \xi_{2i+1}\in[x_{2i},x_{2i+1}],\quad i=0,1,\ldots,n.$$

Then in each segment $[x_{2i},x_{2i+1}]$, we will have two collocation nodes, whereas such nodes in the adjacent segments $[x_{2i+1},x_{2i+2}]$ are absent. We choose the collocation points in each of the segments $[x_{2i},\ x_{2i+1}]$ in the following fashion:

$$\xi_{2i}=x_{2i}+t_1h,\xi_{2i+1}=x_{2i}+t_2h,\quad i=0,1,2,\ldots,n.$$

where t_1 and t_2 are the roots of the second-order Legendre polynomial on the segment [0,1], namely:

$$t_1=\frac{1}{2}-\frac{\sqrt{3}}{6},\quad t_2=\frac{1}{2}+\frac{\sqrt{3}}{6}.$$

Such collocation nodes are called optimal ones. In the first case, when the collocation nodes coincide with the nodes of the mesh $\varDelta$, the spline-collocation method makes it possible to obtain an approximate solution of the problem (2.2.11), (2.2.12) with an accuracy up to $O(h^2)$. Then, by using the last scheme of the arrangement of the collocation points we may reach an accuracy of $O(h^3)$. In some cases it would be reasonable to exclude the unknowns b_{-1},b_{N+1} at the first stage of the problem solution. To do this, we will search for the solution of the boundary-value problem in the form:

$$S(x)=\sum_{i=0}^{N}b_i\phi_i(x),\qquad (2.2.20)$$

where $\phi_i(x)$ are linear combinations of the third-power B-splines satisfying the specified boundary conditions. Thus, the sought spline $S(x)$, which approximates the solution of the boundary-value problem, a priori satisfies exactly the boundary conditions. In this representation of the solution, we have $N+1$ unknown parameters b_i $i = 0, 1, \ldots, N$. For this reason, when using this method, it is necessary to have $N + 1$ collocation nodes.

The functions $\phi_i(x)$ are constructed by allowing for the proper kinds of boundary conditions. Values of $B_3^i(x)$ at the nodes of the mesh Δ are given in Table 2.2. Let us present some variants of the boundary conditions and corresponding functions $\phi_i(x)$:

$$
\begin{aligned}
y(a) &= y(b) = 0(y'(a) \neq 0, y'(b) \neq 0), \\
\phi_0(x) &= -4B_3^{-1}(x) + B_3^0(x),\ \phi_1(x) = B_3^{-1}(x) - \frac{1}{2}B_3^0(x) + B_3^1(x), \\
\phi_i(x) &= B_3^i(x), \quad i = 2, 3, \ldots, N-2, \\
\phi_{N-1}(x) &= B_3^{N-1}(x) - \frac{1}{2}B_3^N(x) + B_3^{N+1}(x),\ \phi_N(x) = B_3^N(x) - 4B_3^{N+1}(x),
\end{aligned} \tag{2.2.21}
$$

$$
\begin{aligned}
y'(a) &= y'(b) = 0\ (y(a) \neq 0,\ y(b) \neq 0), \\
\phi_1(x) &= B_3^{-1}(x) - \frac{1}{2}B_3^0(x) + B_3^1(x), \\
\phi_i(x) &= B_3^i(x), \quad i = 0, 2, 3, \ldots, N-2, N \\
\phi_{N-1}(x) &= B_3^{N-1}(x) - \frac{1}{2}B_3^N(x) + B_3^{N+1}(x),
\end{aligned} \tag{2.2.22}
$$

$$
\begin{aligned}
y(a) &= y'(b) = 0\ (y'(a) \neq 0, y(b) \neq 0), \\
\phi_0(x) &= -4B_3^{-1}(x) + B_3^0(x), \phi_1(x) = B_3^{-1}(x) - \frac{1}{2}B_3^0(x) + B_3^1(x), \\
\phi_i(x) &= B_3^i(x), \quad i = 2, 3, \ldots, N-2, N, \\
\phi_{N-1}(x) &= B_3^{N-1}(x) - \frac{1}{2}B_3^N(x) + B_3^{N+1}(x).
\end{aligned} \tag{2.2.23}
$$

2.2.4 *Reduction of Two-Dimensional Problems to One-Dimensional Ones by Spline-Approximation*

We will now attempt to find the solution $\bar{u}(x, y)$ of the system of linear partial differential equations

$$
\bar{F}\left(x, y, \bar{u}, \frac{\partial \bar{u}}{\partial x}, \frac{\partial \bar{u}}{\partial y}, \frac{\partial^2 \bar{u}}{\partial x^2}, \frac{\partial^2 \bar{u}}{\partial x \partial y}, \frac{\partial^2 \bar{u}}{\partial y^2}, \ldots\right) = 0, \tag{2.2.24}
$$

in the rectangular domain $R\{x_0 \le x \le x_N, y_1 \le y \le y_2\}$ subject to the boundary conditions

$$\overline{G}\left(x, y, \bar{u}, \frac{\partial \bar{u}}{\partial x}, \frac{\partial \bar{u}}{\partial y}\right)\bigg|_{\Gamma} = 0. \tag{2.2.25}$$

Here $\bar{F}$ and $\bar{G}$ are linear vector functions of theirs arguments, Γ is the boundary of the domain R.

We will perform the approximation of the solution of the boundary-value problem (2.2.24) and (2.2.25) in the direction of Ox-axis. For this purpose we construct the mesh $\Delta : x_0 < x_1 < \cdots < x_N$ on the segment $[x_0, x_N]$. We will seek the solution of the boundary-value problem in the form

$$\bar{u}(x, y) = \sum_{i=0}^{N} \bar{u}_i(y)^* \bar{\psi}_i(x), \tag{2.2.26}$$

where $\bar{u}_i(y)$ are to-be-determined vector-functions, i.e., components of the vector-functions $\psi_i(x)$, which are linear combinations of B-splines satisfying the boundary conditions at the sides $x = x_0$ and $x = x_N$ of the rectangle R.

The symbol $\bar{a} * \bar{b}$ stands for a vector, whose components are the product of the corresponding components of the vectors $\bar{a}$ and $\bar{b}$. The B-splines should be chosen in such a way that their powers would exceed the order of the major derivatives of the solution $\bar{u}$ in the system of Eq. (2.2.24).

We now substitute the solution (2.2.26) into the system (2.2.24) and require that its residual will be equal to zero at the straight lines $x = \xi_k,\ k = 0, 1, 2, \ldots, N$ of the rectangle R. These straight lines start from the collocation points (ξ_k, y_1) on the side $y = y_1$. We then obtain the following system of ordinary differential equations of higher order:

$$\begin{aligned} &\bar{F}\Bigg(\xi_k, y, \sum_i \bar{u}_i(y) * \bar{\psi}(\xi_k), \sum_i \bar{u}_i(y) * \frac{d\bar{\psi}_i(\xi)}{dx}, \\ &\sum_i \frac{d\bar{u}_i(y)}{dy} * \bar{\psi}_i(\xi_k), \sum_i \bar{u}_i(y) * \frac{d^2\bar{\psi}_i(\xi_k)}{dx^2}, \\ &\sum_i \frac{d\bar{u}_i(y)}{dy} * \frac{d\bar{\psi}_i(\xi_k)}{dx}, \sum_i \frac{d^2\bar{u}_i(y)}{dy^2} * \bar{\psi}_i(\xi_k), \ldots\Bigg) = 0. \end{aligned} \tag{2.2.27}$$

The boundary conditions (2.2.25) on the sides $y = y_1$ and $y = y_2$ of the rectangle R at the points (ξ_k, y_1), (ξ_k, y_2), $k = 0, 1, \ldots, N$ while observing (2.2.26) can be written as:

$$\begin{aligned} &\bar{G}\Bigg(\xi_k, y_j, \sum_i \bar{u}_i(y_j) * \bar{\psi}_i(\xi_k), \sum_i \bar{u}_i(y_j) * \frac{\mathrm{d}\bar{\psi}_i(\xi_k)}{\mathrm{d}x}, \\ &\sum_i \frac{\mathrm{d}\bar{u}_i(y_j)}{\mathrm{d}y} * \bar{\psi}_i(\xi_k), \ldots\Bigg) = 0, \quad j = 1, 2; \quad k = 0, 1, \ldots, N. \end{aligned} \tag{2.2.28}$$

We will solve the one-dimensional boundary-value problem (2.2.27) and (2.2.28) by using one of the methods of numerical analysis, for example, the method of discrete orthogonalization. To this end, we write this problem preliminary in the normal Cauchy form.

One of the important steps when using the spline-collocation method is constructing combinations of vector functions $\bar{\psi}_i(x)$ that satisfy specified boundary conditions. Earlier we presented some variants of boundary conditions and related linear combinations of third-power B-splines. Later on we will use also fifth-power B-splines. For simplicity, we will assume that the solution of the boundary-value problem is a scalar function $u(x, y)$.

Let us construct the linear combinations of the fifth-power B-splines for some kinds of boundary conditions on sides $x = \text{const}$ of the rectangle R.

To this end, we will use the values $B_5^i(x)$ at nodes of the mesh Δ from Table 2.2 and properties of the B-splines. Consider the following boundary conditions:

$$\begin{aligned} u(x_0, y_j) &= \frac{\partial u(x_0, y_j)}{\partial x} = u(x_N, y_j) = \frac{\partial u(x_N, y_j)}{\partial x} = 0, \\ \psi_0(x) &= \frac{165}{4} B_5^{-2}(x) - \frac{33}{8} B_5^{-1}(x) + B_5^0(x), \\ \psi_1(x) &= B_5^{-1}(x) - \frac{26}{33} B_5^0(x) + B_5^1(x), \\ \psi_2(x) &= B_5^{-2}(x) - \frac{1}{33} B_5^0(x) + B_5^2(x), \\ \psi_i(x) &= B_5^i(x), \quad i = 3, 4, \ldots, N-3, \\ \psi_{N-2}(x) &= B_5^{N-2}(x) - \frac{1}{33} B_5^N(x) + B_5^{N+2}(x), \\ \psi_{N-1}(x) &= B_5^{N-1}(x) - \frac{26}{33} B_5^N(x) + B_5^{N+1}(x), \\ \psi_N(x) &= B_5^N(x) - \frac{33}{8} B_5^{N+1}(x) + \frac{164}{4} B_5^{N+2}(x), \end{aligned} \tag{2.2.29}$$

$$
\begin{aligned}
u(x_0, y_j) &= \frac{\partial^2 u(x_0, y_j)}{\partial x^2} = u(x_N, y_j) = \frac{\partial^2 u(x_N, y_j)}{\partial x^2} = 0, \\
\psi_0(x) &= 12B_5^{-2}(x) - 3B_5^{-1}(x) + B_5^0(x), \\
\psi_1(x) &= -B_5^{-1}(x) + B_5^1(x), \\
\psi_2(x) &= -B_5^{-2}(x) + B_5^2(x), \\
\psi_i(x) &= B_5^i(x), \quad i = 3, 4, \ldots, N-3, \\
\psi_{N-2}(x) &= B_5^{N-2}(x) - B_5^{N+2}(x), \\
\psi_{N-1}(x) &= B_5^{N-1}(x) - B_5^{N+1}(x), \\
\psi_N(x) &= B_5^N(x) - 3B_5^{N+1}(x) + 12B_5^{N+2}(x),
\end{aligned}
\tag{2.2.30}
$$

$$
\begin{aligned}
&\frac{\partial u(x_0, y_j)}{\partial x} = \frac{\partial^3 u(x_0, y_j)}{\partial x^3} = \frac{\partial u(x_N, y_j)}{\partial x} = \frac{\partial^3 u(x_N, y_j)}{\partial x^3} = 0, \\
&\psi_1(x) = B_5^{-1}(x) + B_5^1(x), \ \psi_2(x) = B_5^{-2}(x) + B_5^2(x), \\
&\psi_i(x) = B_5^i(x), \quad i = 0, 3, 4, \ldots, N-3, N \\
&\psi_{N-2}(x) = B_5^{N-2}(x) + B_5^{N+2}(x), \ \psi_{N-1}(x) = B_5^{N-1}(x) + B_5^{N+1}(x),
\end{aligned}
\tag{2.2.31}
$$

By combining these conditions on different sides $x = \text{const.}$ of the rectangle, we can extend the number of variants of boundary conditions.

The spline collocation method was applied for the first time to the solution of state-strain shells problems in Grigorenko and Berenov [8].

2.3 Discrete Fourier Series Method

We will now consider an approach for solving two-dimensional boundary-value problems described by partial differential equations with variable coefficients in two coordinate directions, characterized by a stress-strain state of linear elastic bodies, and subjected to different loads under certain boundary conditions. This approach is based on the use of the discrete Fourier series which reduces the two-dimensional boundary-value problem to a one-dimensional (Grigorenko [7]).

The stress-strain state of the elastic body is described by a system of partial differential equations in the form

$$
\frac{\partial Z_i}{\partial \alpha} = \Phi_i\left(\alpha, \beta, \frac{\partial^k Z_j}{\partial \beta^k}\right) + f_i(\alpha, \beta), \quad i, j, k = \overline{1, l}, \tag{2.3.1}
$$

where $Z_i = Z_i(\alpha, \beta)$, $\alpha_1 \leq \alpha \leq \alpha_2, \beta_1 \leq \beta \leq \beta_2$ are the sought governing functions; Φ_i are linear functions with respect to their arguments, $f_i(\alpha, \beta)$ are the right-hand sides, and $\alpha O \beta$ refers to an orthogonal curvilinear coordinate system.

This system for unclosed elastic bodies (plates, shallow shells and other) is complemented by boundary conditions at the contours $\alpha = \text{const.}$ and $\beta = \text{const.}$ For bodies bound in one coordinate direction the boundary conditions are replaced by periodicity conditions in this direction.

Due to the periodicity of all the sought functions the boundary-value problem for the system of Eq. (2.3.1) for closed elastic bodies, for example in the $O\beta$ direction, makes it possible to present the solutions in form of Fourier series with respect to the coordinate β. However, in this case all terms in the equations with associated coefficients do not prevent separation of variables in this direction. In the more simple problems, the variables can be separated by representing all the functions in form of Fourier series. But in many cases the system of differential Eq. (2.3.1) includes terms with coefficients characterizing geometrical and mechanical parameters that make it impossible to separate the variables and to present the sought functions in form of Fourier series. To overcome these difficulties, additional functions, which are expressed in terms of the governing functions and their derivatives, are introduced. Then we arrive at the following governing system of equations:

$$\frac{\partial Z_i}{\partial \alpha} = F_i\left(\alpha, \beta, \frac{\partial^k Z_j}{\partial \beta^k}, \varphi_r^p\right) + f_i(\alpha, \beta), (i, j, k) = \overline{1, l},\; r = \overline{1, R},\; p = \overline{1, P},$$

where:

$$\varphi_r^p = \varphi_r^p\left(\alpha, \beta, \frac{\partial^s Z_i}{\partial \alpha^s}, \frac{\partial^t Z_i}{\partial \beta^t}\right), (s, t) \le l). \tag{2.3.2}$$

Note that the system of differential equations (2.3.2) contains, except for the governing functions Z_i, additional functions φ_r^p. Because of this the general number of unknown functions must not exceed the order of the system of equations. This fact should be taken into account when solving the boundary-value problem. To solve the original boundary-value problem, we expand all the functions in (2.3.2) into Fourier series with respect to the coordinate β:

$$\begin{aligned} \tilde{X}(\alpha, \beta) &= \sum_{m=0}^{M} \tilde{X}_m(\alpha) \cos \lambda_m \beta, \tilde{Y}(\alpha, \beta) \\ &= \sum_{m=0}^{M} \tilde{Y}_m(\alpha) \sin \lambda_m \beta, \lambda_m = 2\pi m/T, \end{aligned} \tag{2.3.3}$$

where $\tilde{X}$ and $\tilde{Y}$ are the governing and additional functions of system (2.3.2), T is the period.

After substituting the series (2.3.3) in (2.3.2), separation of the variables, and some transformations for the amplitude values of the series (2.3.3), we arrive at the following coupled system of ordinary differential equations:

$$\frac{dZ_{im}}{d\alpha} = F_{im}(\alpha, Z_{im}, \varphi^{p}_{rm}) + f_{im}(\alpha), i = \overline{1, l}, m = \overline{0, M}, r = \overline{1, R}, p = \overline{1, P}. \quad (2.3.4)$$

By performing the same transformations with the boundary conditions we determine boundary conditions for the amplitude values of the functions Z_{im} at the ends of the interval $\alpha_1 \leq \alpha \leq \alpha_2$. In order to solve the boundary-value problem for the system of Eq. (2.3.4) when subjected to the associated boundary conditions, we will use the stable numerical discrete-orthogonalization method (cf., Sect. 2.1). In many cases the system (2.3.4) will be stiff due to inhomogeneity of mechanical and geometrical parameters as well as to the load applied in the original problem. The discrete-orthogonalization method, which is used to solve the one-dimensional boundary-value problem, is based on the orthogonalization of the vector-solutions of the Cauchy problems at a finite number of points of the interval within which the argument varies. The system of Eq. (2.3.4), along with the amplitude values of the governing functions, still includes the amplitude values of additional functions, which should be determined separately. By using the amplitude values of the governing functions simultaneously for all harmonics at each step for the fixed value of α, when integrating the system (2.3.4) by means of the discrete-orthogonalization method, we calculate the amplitude values of the additional functions at some points of the interval $\beta_1 \leq \beta \leq \beta_2$. Then we set up the Fourier series for the functions specified on the discrete ensemble of points. By increasing the number of points, at which the values of the additional functions are calculated, the discrete Fourier series differ progressively smaller from the exact Fourier series and, consequently, guarantee a highly accurate results. By using the Runge scheme, we determine the coefficients of these series, substitute them into system (2.3.4), and continue its integration. The boundary conditions at the ends of the interval $\alpha_1 \leq \alpha \leq \alpha_2$ should be satisfied.

The general scheme of the calculations given above can be varied depending on the specific form of the right-hand sides of the equations in system (2.3.1). During the solution of most cases of applied problems, only some first terms of discrete Fourier series are used: The Fourier coefficients decrease rapidly which causes distant harmonics to do the same. It is known that the rate at which the Fourier coefficients decrease, is important for the accuracy of approximations in calculations involving Fourier series. This situation is attributed to differential properties of the function extended to the interval $(-\infty, \infty)$. The results obtained with the above approach are presented below.

We assume that the function $y(x)$ is given for an ensemble of points, i.e., $y(x_i) = y_i$ for $x_i = i2\pi/k,\ i = 0, 1, 2, \ldots, k - 1$. We want to construct the Fourier series for the function $f(x)$ prescribed at the discrete ensemble of the points $x_i, i = \overline{0, k-1}$. Assume that this series takes the form

$$y(x) = a_0 + \sum_{m=1}^{n} (a_m \cos mx + b_m \sin mx), \quad n \leq k/2, \tag{2.3.5}$$

where the coefficients a_0, a_m, and b_m are determined as follows:

$$\begin{aligned} a_0 &= \frac{1}{k}\sum_{i=0}^{k-1} y_i, a_m = \frac{2}{k}\sum_{i=0}^{k-1} y_i \cos m\frac{2\pi i}{k}, \\ b_m &= \frac{2}{k}\sum_{i=1}^{k-1} y_1 \sin m\frac{2\pi i}{k}, \quad m \leq k/2. \end{aligned} \tag{2.3.6}$$

We define relationships relating approximate values of the coefficients of the Fourier series with the exact values of the same coefficients. Consider the double differentiable function $y = f(x)$ given analytically on the interval $[0, 2\pi]$. The exact Fourier series for this function is:

$$y(x) = A_0 + \sum_{j=1}^{\infty} A_j \cos jx + \sum_{j=1}^{\infty} B_j \sin jx. \tag{2.3.7}$$

Here capital letters denote exact values of the coefficients. By putting $x_i = i2\pi/k,\ i = 0, 1, 2, \ldots, k-1$, we calculate those values of the function $y_i = y(x_i)$ which appear in Eq. (1.6). After substituting these values of y_i into (2.3.6) and some transformations we obtain:

$$\begin{aligned} a_m &= A_m + A_{k-m} + A_{k+m} + A_{2k-m} + \cdots, \\ b_m &= B_m - B_{k-m} + B_{k+m} - B_{2k-m} + \cdots, m \leq k/2. \end{aligned}$$

In particular we have:

1. at $k = 12$:

$$\begin{aligned} a_0 &= A_0 + A_{12} + \cdots, a_1 = A_1 + A_{11} + \cdots, a_2 = A_2 + A_{10} + \cdots, a_3 \\ &= A_3 + A_9 + \cdots, \end{aligned}$$

2. at $k = 24$:

$$\begin{aligned} &a_0 = A_0 + A_{24} + \cdots, a_1 = A_1 + A_{23} + \cdots, a_2 = A_2 + A_{22} + \cdots, a_3 = A_3 + A_{21} + \cdots, \\ &a_4 = A_4 + A_{20} + \cdots, a_5 = A_5 + A_{19} + \cdots, a_6 = A_6 + A_{18} + \cdots, a_7 = A_7 + A_{17} + \cdots, \\ &a_8 = A_8 + A_{16} + \cdots \end{aligned}$$

From these equalities it follows that for $k = 12$ only 2 or 3 harmonics should be taken into account, while for $k = 24$ the sufficient accuracy is reached for the first 7 or 8 harmonics.

References

1. Alberg JH, Nielson E, Walsh J (1967) Theory of splines and their applications. Academic Press, New York
2. Bellman R, Kalaba R (1965) Quasilinearization and nonlinear boundary-value problems. Elsevier
3. Collatz L (1951) Numerische Behandlung von Differentialgleichungen. Berlin-Göttingen-Heidelberg
4. Grigorenko AY (2005) Numerical analysis of stationary dynamic processes in anisotropic inhomogeneous cylinders. Int Appl Mech 41(8):831–866
5. Grigorenko AY, Efimova TL (2005) Spline-approximation method applied to solve natural-vibration problems for rectangular plates of varying thickness. Int Appl Mech 41 (10):1161–1169
6. Grigorenko YM (1984) Solution of problems in the theory of shells by numerical-analysis methods. Int Appl Mech 20(10):881–897
7. Grigorenko YM (2009) Using discrete Fourier series to solve boundary-value stress problems for elastic bodies with complex geometry and structures. Int Appl Mech 45(5):470–513
8. Grigorenko YM, Berenov MN (1988) Numerical solution of problems in the statics of flattened shells on the basis of the spline collocation method. Int Appl Mech 24(5):458–463
9. Grigorenko YM, Grigorenko AY (2013) Static and dynamic problems for anisotropic inhomogeneous shells with variable parameters and their numerical solution (review). Int Appl Mech 49(2):123–193
10. Grigorenko YM, Grigorenko AY, Efimova TL (2008) Spline-based investigation of natural vibrations of orthotropic rectangular plates of variable thickness within classical and refined theories. J Mech Struct 3(5):929–952
11. Grigorenko YM, Vasilenko AT (1997) Solution of problems and analysis of the stress-strain state of nonuniform anisotropic shells (survey). Int Appl Mech 33(11):851–880
12. Grigorenko YM, Vasilenko AT (2002) Some approaches to the solution of problems on thin shells with variable geometrical and mechanical parameters. Int Appl Mech 38(11):1309–1341

Chapter 3
Some Solutions for Anisotropic Heterogeneous Shells Based on Classical Model

Abstract Results for stationary deformation of anisotropic inhomogeneous shells of various classes are presented by using classical Kirchhoff-Love theory and the numerical approaches outlined in Chap. 2 of this book. The stress-strain problems for shallow, noncircular cylindrical shells and shells of revolution are solved. Various types of boundary conditions and loadings are considered. Distributions of stress and displacement fields in shells of the aforementioned type are analyzed for various geometrical and mechanical parameters. The practically important stress problem of a high-pressure glass-reinforced balloon is solved. Dynamical characteristics of an inhomogeneous orthotropic plate under various boundary conditions are studied. The problem of free vibrations of a circumferential inhomogeneous truncated conical shell is solved. The effect of variation in thicknesses, mechanical parameters, and boundary conditions on the behavior of natural frequencies and vibration modes of a plate and cone is analyzed. Much attention is given to the validation of the reliability of the results obtained by numerical calculations.

Keywords Stress-strain state · Free vibrations · Anisotropy · Heterogeneity · Rectangular plates · Shallow · Spherical · Conical and noncircular shells

3.1 Stress-Strain State of Shallow Shells

We consider a multilayer shallow rectangular in-plane shell (Fig. 3.1), symmetrical about the mid-surface of the structure and composed of an odd number of orthotropic layers with variable-thickness (Grigorenko and Vasilenko [9], Librescu and Hause [14], Librescu and Schmidt [15], Noor and Burton [16], Ramm [17]). It is assumed that the layers operate jointly without separation and sliding. As initial model, we adopt the mathematical one based on the works of Donnell [2] and Vlasov [20]. The relevant equations for this model were presented in Chap. 1.

The governing relations of the model can be written as follows; the expressions for the strains:

A.Ya. Grigorenko et al., *Recent Developments in Anisotropic Heterogeneous Shell Theory*, SpringerBriefs in Continuum Mechanics,
DOI 10.1007/978-981-10-0353-0_3

Fig. 3.1 Multilayer shallow rectangular in-plane shell

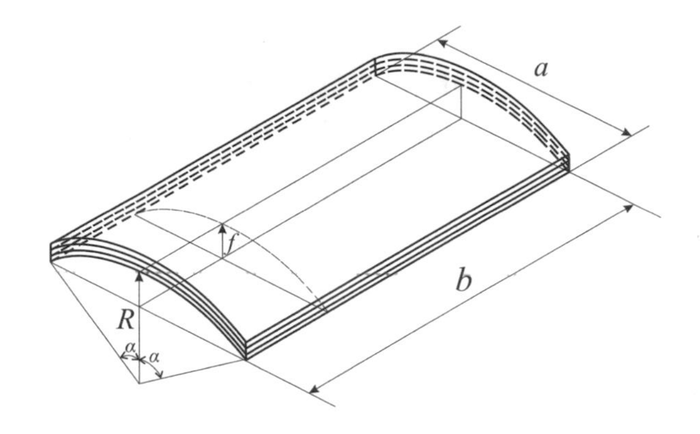

$$\varepsilon_x = \frac{\partial u}{\partial x} + \frac{w}{R_x}, \quad k_x = -\frac{\partial^2 w}{\partial x^2},$$
$$\varepsilon_y = \frac{\partial v}{\partial y} + \frac{w}{R_y}, \quad k_y = -\frac{\partial^2 w}{\partial y^2}, \tag{3.1.1}$$
$$\varepsilon_{xy} = \frac{\partial u}{\partial y} + \frac{\partial v}{\partial x}, \quad k_{xy} = -\frac{\partial^2 w}{\partial x \partial y};$$

the equilibrium equations:

$$\frac{\partial N_x}{\partial x} + \frac{\partial S}{\partial y} = 0, \quad \frac{\partial N_y}{\partial y} + \frac{\partial S}{\partial x} = 0,$$
$$\frac{\partial M_x}{\partial x} + \frac{\partial H}{\partial y} = Q_x, \quad \frac{\partial M_y}{\partial y} + \frac{\partial H}{\partial x} = Q_y, \tag{3.1.2}$$
$$\frac{\partial Q_x}{\partial x} + \frac{\partial Q_y}{\partial y} - \frac{N_x}{R_x} - \frac{N_y}{R_y} + q_\gamma = 0;$$

and the elasticity relations:

$$N_x = C_{11}\varepsilon_x + C_{12}\varepsilon_y, \quad N_y = C_{12}\varepsilon_x + C_{22}\varepsilon_y, \quad S = C_{66}\varepsilon_{xy},$$
$$M_x = D_{11}k_x + D_{12}k_y, \quad M_y = D_{12}k_x + D_{22}k_y, \quad H = 2D_{66}k_{xy}. \tag{3.1.3}$$

The coefficients appearing in (3.1.3) are determined by:

$$C_{mp} = \sum_{i=1}^{2n+1} \int_{\gamma_{i-1}}^{\gamma_i} B_{mp} \mathrm{d}\gamma, \quad D_{mp} = \sum_{i=1}^{2n+1} \int_{\gamma_{i-1}}^{\gamma_i} B_{mp}\gamma^2 \mathrm{d}\gamma, \quad m, p = 1, 2, 6, \tag{3.1.4}$$

where the expressions B^i_{mp} for the ith layer of isotropic material are (for simplicity the index i is omitted):

$$B_{11} = \frac{E_x}{1 - \nu_x \nu_y}, \quad B_{12} = \frac{\nu_x E_y}{1 - \nu_x \nu_y} = \frac{\nu_y E_x}{1 - \nu_x \nu_y}, \\ B_{22} = \frac{E_y}{1 - \nu_x \nu_y}, \quad B_{66} = G_{xy}. \tag{3.1.5}$$

In the case of an isotropic layer, the coefficients have the form:

$$B_{11} = B_{22} = \frac{E}{1 - \nu^2}, \quad B_{12} = \frac{\nu E}{1 - \nu^2}, \quad B_{66} = \frac{\nu E}{2(1 + \nu)}. \tag{3.1.6}$$

The following notations are used in context with the relations (3.1.1)–(3.1.6): x, y are the coordinates ($0 \le x \le a, \quad 0 \le y \le b$); u, v, w, are the displacements of the mid-surface; $\varepsilon_x, \varepsilon_y, \varepsilon_{xy}, k_x, k_y, k_{xy}$ are the tangential and normal strains; N_x, N_y, S, Q_x, Q_y are the tangential and normal forces; M_x, M_y, and H are the bending and twisting moments; E_x, E_y are the elastic moduli in the directions of coordinate axes; G_{xy} is the shear modulus at a tangential plane; ν_x and ν_y are Poisson's ratios; R_x and R_y are radii of curvature.

After some transformations, we obtain from Eqs. (3.1.1)–(3.1.6) the system of governing equations for displacements with variable coefficients:

$$\begin{aligned}
&C_{11}\frac{\partial^2 u}{\partial x^2} + C_{66}\frac{\partial^2 u}{\partial y^2} + (C_{12} + C_{66})\frac{\partial^2 v}{\partial x \partial y} + \frac{\partial C_{11}}{\partial x}\frac{\partial u}{\partial x} + \frac{\partial C_{66}}{\partial y}\frac{\partial u}{\partial y} \\
&\quad + \frac{\partial C_{66}}{\partial y}\frac{\partial v}{\partial x} + \frac{\partial C_{12}}{\partial x}\frac{\partial v}{\partial y} + \frac{\partial}{\partial x}\left(\frac{C_{11}}{R_x} + \frac{C_{12}}{R_y}\right)w + \left(\frac{C_{11}}{R_x} + \frac{C_{12}}{R_y}\right)\frac{\partial w}{\partial x} = 0, \\
&C_{66}\frac{\partial^2 v}{\partial x^2} + C_{22}\frac{\partial^2 v}{\partial y^2} + (C_{12} + C_{66})\frac{\partial^2 u}{\partial x \partial y} + \frac{\partial C_{22}}{\partial y}\frac{\partial v}{\partial y} \\
&\quad + \frac{\partial C_{66}}{\partial x}\frac{\partial v}{\partial x} + \frac{\partial C_{66}}{\partial x}\frac{\partial u}{\partial y} + \frac{\partial C_{12}}{\partial y}\frac{\partial u}{\partial x} + \frac{\partial}{\partial y}\left(\frac{C_{12}}{R_x} + \frac{C_{22}}{R_y}\right)w + \left(\frac{C_{12}}{R_x} + \frac{C_{12}}{R_y}\right)\frac{\partial w}{\partial y} = 0, \\
&D_{11}\frac{\partial^4 w}{\partial x^4} + D_{22}\frac{\partial^4 w}{\partial y^4} + (2D_{12} + 4D_{66})\frac{\partial^4 w}{\partial x^2 \partial y^2} + 2\frac{\partial D_{11}}{\partial x}\frac{\partial^3 w}{\partial x^3} \\
&\quad + 2\frac{\partial D_{22}}{\partial y}\frac{\partial^3 w}{\partial y^3} + \left(2\frac{\partial D_{12}}{\partial y} + 4\frac{\partial D_{66}}{\partial y}\right)\frac{\partial^3 w}{\partial x^2 \partial y} + \left(2\frac{\partial D_{12}}{\partial x} + 4\frac{\partial D_{66}}{\partial x}\right)\frac{\partial^3 w}{\partial x \partial y^2} \\
&\quad + \left(\frac{\partial^2 D_{11}}{\partial x^2} + \frac{\partial^2 D_{12}}{\partial y^2}\right)\frac{\partial^2 w}{\partial x^2} + \left(\frac{\partial^2 D_{12}}{\partial x^2} + \frac{\partial^2 D_{22}}{\partial y^2}\right)\frac{\partial^2 w}{\partial y^2} + 4\frac{\partial^2 D_{66}}{\partial x \partial y}\frac{\partial^2 w}{\partial x \partial y} \\
&\quad + \frac{C_{11}}{R_x}\frac{\partial u}{\partial x} + \frac{C_{12}}{R_y}\frac{\partial v}{\partial y} + \frac{1}{R_x}\left(\frac{C_{11}}{R_x} + \frac{C_{12}}{R_y}\right)w + \frac{C_{12}}{R_y}\frac{\partial u}{\partial x} + \frac{C_{22}}{R_y}\frac{\partial v}{\partial y} + \frac{1}{R_y}\left(\frac{C_{12}}{R_x} + \frac{C_{22}}{R_y}\right)w = q_y.
\end{aligned} \tag{3.1.7}$$

After adding to the system of equations (3.1.7) the boundary conditions on the shell contours $x = 0,\ x = a$ and $y = 0,\ y = b$, we arrive at a two-dimensional boundary-value problem.

We now present the system of equations (3.1.7) as follows:

$$\begin{aligned}
\frac{\partial^2 u}{\partial x^2} &= f_1\left(\frac{\partial^2 u}{\partial y^2}, \frac{\partial^2 v}{\partial x \partial y}, \frac{\partial u}{\partial x}, \frac{\partial u}{\partial y}, \frac{\partial v}{\partial x}, \frac{\partial v}{\partial y}, \frac{\partial w}{\partial x}, w\right),\\
\frac{\partial^2 v}{\partial x^2} &= f_2\left(\frac{\partial^2 v}{\partial x^2}, \frac{\partial^2 u}{\partial x \partial y}, \frac{\partial v}{\partial y}, \frac{\partial v}{\partial x}, \frac{\partial u}{\partial y}, \frac{\partial u}{\partial x}, \frac{\partial w}{\partial y}, w\right),\\
\frac{\partial^4 w}{\partial x^4} &= f_3\left(\frac{\partial^4 w}{\partial y^4}, \frac{\partial^4 w}{\partial x^2 \partial y^2}, \frac{\partial^3 w}{\partial x^3}, \frac{\partial^3 w}{\partial y^3}, \frac{\partial^3 w}{\partial x^2 \partial y}, \frac{\partial^3 w}{\partial x \partial y^2},\right.\\
&\qquad \left.\frac{\partial^2 w}{\partial x^2}, \frac{\partial^2 w}{\partial y^2}, \frac{\partial^2 w}{\partial x \partial y}, \frac{\partial u}{\partial x}, \frac{\partial v}{\partial y}, \frac{\partial u}{\partial x}, \frac{\partial v}{\partial y}, w, q_\gamma\right).
\end{aligned} \tag{3.1.8}$$

In order to reduce the dimensionality of the system of partial differential equations (3.1.8), we use the spline-approximation of the solution in one coordinate direction. We will seek the solution of the boundary-value problem for the system of equations (3.1.8) in the form:

$$\begin{aligned}
u(x,y) &= \sum_{i=0}^{N} u_i(x)\phi_i(y), \quad v(x,y) = \sum_{i=0}^{N} v_i(x)\phi_i(y),\\
w(x,y) &= \sum_{i=0}^{N} w_i(x)\psi_i(y),
\end{aligned} \tag{3.1.9}$$

where u_i, v_i, w_i, $i = \overline{0,N}$ are the sought functions; ϕ_i and ψ_i, $i = \overline{0,N}; N \geq 6$ are functions constructed by using B-splines of the third and fifth power, respectively. These splines make it possible to compose their linear combinations such that the different conditions on the shell contours $y = 0, y = b = 0$ are satisfied. In this case, the functions $\phi_i(y)$ and $\psi_i(y)$ can be represented in terms of B-splines as follows:

$$\begin{aligned}
\phi_0(y) &= \alpha_{11}B_3^{-1} + \alpha_{12}B_3^0,\\
\phi_1(y) &= B_3^{-1} + \alpha_{21}B_3^0 + \alpha_{22}B_3^1,\\
\phi_i(y) &= B_3^i, \quad i = \overline{2, N-2},\\
\phi_{N-1}(y) &= B_3^{N+1} + \beta_{21}B_3^N + \beta_{22}B_3^{N-1},\\
\phi_N(y) &= \beta_{11}B_3^{N+1} + \beta_{21}B_3^N,\\
\psi_0 &= \gamma_{11}B_5^{-2} + \gamma_{12}B_5^{-1} + B_5^0,\\
\psi_1 &= \gamma_{21}B_5^{-1} + \gamma_{22}B_5^0 + B_5^1,\\
\psi_2 &= \gamma_{31}B_5^{-2} + \gamma_{32}B_5^0 + B_5^2,\\
\psi_i &= B_5^i, \quad i = \overline{3, N-3},\\
\psi_{N-2} &= \delta_{31}B_5^{N+2} + \delta_{32}B_5^N + B_5^{N-2},\\
\psi_{N-1} &= \delta_{21}B_5^{N+1} + \delta_{22}B_5^N + B_5^{N-1},\\
\psi_N &= \delta_{11}B_5^{N+2} + \delta_{12}B_5^{N+1} + B_5^N,
\end{aligned} \tag{3.1.10}$$

where B_3^i, $i = \overline{-1, N+1}$ are cubic B-splines, B_5^i $i = \overline{-2, N+2}$ are fifth-power splines, constructed at a uniform mesh, $\alpha, \beta, \gamma, \delta$ are constants to-be-determined depending on the boundary conditions specified on the contours.

Having constructed linear combinations of B-splines in the form of functions $\phi_i(y)$ and $\psi_i(y)$ so that certain boundary conditions on the contours $y = 0$ and $y = b$ are satisfied, we now substitute the expressions (3.1.9) into the governing equations (3.1.8) and require them to be satisfied at the collocation points y_k, $k = \overline{0, N}$. Then we obtain a system of $3(N + 1)$ ordinary differential equations. The boundary conditions on the contours $x = 0$ and $x = a$ are handled similarly. The obtained system of differential equations together with the related boundary conditions form a two-point boundary-value problem on the interval $0 \le x \le a$.

After introducing the following notations:

$$\bar{Y} = \{\bar{y}_1, \bar{y}_2, \ldots, \bar{y}_8\}^{\mathrm{T}} = \{\bar{u}, \bar{u}', \bar{v}, \bar{v}', \bar{w}, \bar{w}', \bar{w}'', \bar{w}'''\}^{\mathrm{T}}, \quad (3.1.11)$$

where $\bar{y}_m = \{y_{m_0}, y_{m_1}, \ldots, y_{m_N}\}^T$, $m = \overline{1, 8}$, the boundary-value problem becomes:

$$\frac{\mathrm{d}\bar{Y}}{\mathrm{d}x} = A(x)\bar{Y} + \bar{g}(x), \quad 0 \le x \le a, \quad (3.1.12)$$

with:

$$B_1\bar{Y}(0) = \bar{b}_1, B_2\bar{Y}(a) = \bar{b}_2. \quad (3.1.13)$$

In order to solve the boundary-value problem for the system of equations (3.1.12) together with the boundary conditions (3.1.13), we use the stable numerical method of discrete orthogonalization. By substituting the defined values of functions $u_i(a), v_i(a), w_i(a)$, $i = \overline{0, N}$ into the expressions (3.1.9), we obtain a solution of a two-dimensional problem for the displacements which is used to determine all parameters of the stress-strain state of the shell.

In order to illustrate the approach, we present the solution of the problem for a shallow cylindrical shell of constant thickness under the distributed transverse load q. We assume that the shell is rigidly fixed along the edges $y = 0, y = b$ and hinged at edges $x = 0$, $x = a$, i.e., the boundary conditions have the following form:

$$u = v = w = \frac{\partial w}{\partial y} = 0, \quad y = 0, \quad y = b, \quad (3.1.14)$$

$$\frac{\partial u}{\partial x} = v = w = \frac{\partial^2 w}{\partial x^2} = 0, \quad x = 0, \quad x = a. \quad (3.1.15)$$

In this case, we choose functions ϕ_i and $\psi_i, i = \overline{0, N}$ such that the conditions (3.1.14) are satisfied. For the solution of the problem, the following data was used:

Table 3.1 Refined solution

Deflection and its nth derivatives	Solution	y				
		0.00	1.25	2.50	3.75	5.00
wq_0/E	In splines N = 7	0.000	120.8	337.9	512.2	576.8
	N = 9	0.000	121.2	338.6	512.7	577.2
	Exact	0.000	121.4	338.8	512.9	577.4
$\frac{q_0}{E}\frac{\partial^2 w}{\partial x^2}$	In splines N = 7	0.000	−8.279	−23.17	−35.10	−39.53
	N = 9	0.000	−8.309	−23.20	−35.14	−39.53
	Exact	0.000	−8.323	−23.22	−35.15	−39.57
$\frac{q_0}{E}\frac{\partial^2 w}{\partial y^2}$	In splines N = 7	216.5	55.63	−31.17	−72.67	−84.57
	N = 9	217.3	55.39	−31.43	−72.51	−84.55
	Exact	217.8	55.18	−31.48	−72.53	−84.46

$$R_2 = 13, a = 12, b = 10, h = 0.4, v = 0.3, q = q_0 \sin\frac{\pi x}{12}, q_0 = \text{const.}$$

Results based on the refined solution and by the proposed method are presented in Tables 3.1 and 3.2, respectively, for $N = 7$ and $N = 9$ at sections $x = 6$ and $y = 5$.

The refined solution is obtained by separation of variables along the Ox-axis and by solving the ordinary differential equations with high accuracy. From the data given in the tables, we may conclude that the proposed approach makes it possible to solve problems with sufficient accuracy.

Table 3.2 Refined solution

Deflection and it nth derivatives	Solution	x				
		0.0	1.5	3.0	4.5	6.0
wq_0/E	In splines N = 7	0.000	220.7	407.8	532.9	576.8
	N = 9	0.000	220.9	408.1	533.3	577.2
	Exact	0.000	221.0	408.3	533.4	577.4
$\frac{q_0}{E}\frac{\partial^2 w}{\partial x^2}$	In splines N = 7	0.000	−15.13	−27.95	−36.52	−39.53
	N = 9	0.000	−15.13	−27.96	−36.54	−39.55
	Exact	0.000	−15.14	−27.98	−36.56	−39.57
$\frac{q_0}{E}\frac{\partial^2 w}{\partial y^2}$	In splines N = 7	0.000	−32.36	−59.79	−78.13	−84.57
	N = 9	0.000	−32.35	−59.79	−78.12	−84.55
	Exact	0.000	−32.32	−59.72	−78.03	−84.46

By using the spline-collocation method and by special choice of the nodal points, we can sufficiently rise the accuracy. Here the collocation points on each subinterval were determined by the zeros of the quadratic Legendre polynomial.

Based on the approach developed to solving problems of the given class, we will analyze the stress-strain state of sandwich shallow shells, whose middle layer is orthotropic and whose outer and inner layers are isotropic. The shells are loaded by a normal force $q_y = q_0 = \text{const.}$ and their sides are rigidly fixed.

By assuming that the elastic and shear moduli are given by $E_x = E$, $E_y = \mu E$, $G_{xy} = \lambda E$, respectively, and that Poisson's ratio is represented by v_x, we consider five variants of elastic constants for the following sets of parameters:

$$\begin{array}{llll} (1) & \mu = 2; & \lambda = 0.3; & v_x = 0.075; \\ (2) & \mu = 1.35; & \lambda = 0.215; & v_x = 0.122; \\ (3) & \mu = 1; & \lambda = 0.385; & v_x = 0.3; \\ (4) & \mu = 0.741; & \lambda = 0.159; & v_x = 0.165; \\ (5) & \mu = 0.5; & \lambda = 0.125; & v_x = 0.15. \end{array} \tag{3.1.16}$$

The values of the elastic constants for the third variant correspond to an isotropic material. The thickness of the middle layer is $h_{\text{orth}} = 0.4\ h$, and of the outer and inner layers we put $h_{\text{isotr}} = 0.3\ h$. The overall thickness of the shell is $h_{\text{orth}} + 2\ h_{\text{isotr}} = h$. The value of the rise is:

$$f = R - \sqrt{R^2 - \Delta^2}, \tag{3.1.17}$$

where R is the radius of curvature, and Δ is the half-length of the side along one of the coordinate directions.

The problems were solved by using the following initial data:

$$\begin{array}{llll} (1) & f = 0.5 & R_x = 36.25; & R_y = 25.25; \\ (2) & f = 1; & R_x = 18.50; & R_y = 13; \\ (3) & f = 2; & R_x = 10; & R_y = 7.25. \end{array} \tag{3.1.18}$$

The sizes of the shell foundation and thickness are $a = 12$; $b = 10$; $h = 0.4$ (the case $f = 0$ corresponds to a plate).

The spline-approximation is carried out in the Oy-axis direction, in order to satisfy the boundary conditions at $y = 0$ and $y = b$ in the expressions (3.1.10). Moreover, we use the following values for the coefficients:

$$\begin{array}{llll} \alpha_{11} = \beta_{11} = -4, & \alpha_{12} = \beta_{12} = 1, & \alpha_{21} = \beta_{21} = -0.5, & \alpha_{22} = \beta_{22} = 1, \\ \gamma_{11} = \delta_{11} = 165/4, & \gamma_{12} = \delta_{12} = -33/8, & \gamma_{21} = \delta_{21} = 1, & \\ \gamma_{22} = \delta_{22} = -26/33, & \gamma_{31} = \delta_{31} = 1, & \gamma_{32} = \delta_{32} = -1/33 & \end{array}$$

We will now analyze the distribution of displacements and stresses for the five variants of elastic constants shown in Eq. (3.1.16) for the middle layer under different values of the rise of the shell (3.1.18).

Table 3.3 summarizes the distributions of displacements w in y at $x = 6$. Table 3.4 does the same for x at $y = 5$ depending on the variation in the orthotropy parameters and the degree of shallowness. The numbers 1, 2, 3, 4, 5 stand for the orthotropy variants for the middle layer. By virtue of symmetry of the problems in x and y the tables shows displacements only up to the lines of symmetry.

From the data presented in Tables 2.6 and 2.7 one can see how the displacements for different values of the rise change with varying mechanical characteristics of the orthotropic material of the middle layer.

The value of maximum deflection for variant 5 is used for scaling the value of deflection for variant 1 at $f = 0$; 0.5; 1; 2, respectively, resulting in factors of 1.05, 1.27, 1.36, and 1.39 times. If the rise increases, the maximum deflection decreases compared to the plate for variant 1 by 2.96, 9.31, 32.7 times, for variant 3 by 2.76, 8.38, 29.8 times, and for variant 5 by 2.47, 7.11, and 24.9 times.

The magnitudes of the stresses σ_y at $x = 6$, $y = 0$ are presented in Table 3.5, where $\sigma^{\pm}_{b,r}$, $\sigma^{\pm}_{b,m}$, $\sigma^{\pm}_{b,l}$ are the stresses on the lateral surfaces of the upper isotropic, the middle orthotropic, and the lower isotropic layers, respectively. The stresses σ_y

Table 3.3 Distributions of displacements w in y at $x = 6$

f	y	Ew/q_0 $(x = 6)$				
		1	2	3	4	5
0	1	410.2	420.5	422.6	431.1	435.7
	2	1235	1264	1268	1294	1307
	3	2066	2114	2118	2161	2183
	4	2052	2712	2715	2771	2799
	5	2861	2925	2928	2989	3018
0.5	1	144.1	158.3	160.9	176.0	155.1
	2	427.3	468.5	474.5	519.6	546.2
	3	705.5	772.1	779.7	854.5	897.1
	4	896.9	980.5	988.4	1084	1138
	5	964.2	1054	1062	1164	1222
1.0	1	51.7	58.6	60.2	67.9	73.0
	2	147.3	166.3	169.5	191.7	205.6
	3	234.0	263.5	267.1	302.6	324.1
	4	289.6	325.5	328.7	372.9	399.0
	5	308.3	346.3	349.2	396.4	424.1
2.0	1	19.4	22.0	22.6	25.8	28.1
	2	50.1	56.5	57.5	65.8	71.2
	3	72.4	81.8	82.6	94.7	102
	4	84.0	94.5	94.9	109	117
	5	87.3	98.1	98.3	113	121

Table 3.4 Distributions of displacements w in x at $y = 5$

f	x	Ew/q_0 $(y = 5)$				
		1	2	3	4	5
0	1.2	475.9	484.1	484.8	492.6	495.8
	2.4	1355	1381	1382	1407	1419
	3.6	2157	2201	2203	2247	2267
	4.8	2682	2740	2743	2799	2826
	6.0	2861	2925	2928	2989	3018
0.5	1.2	195.1	206.3	206.4	220.7	228.5
	2.4	517.3	553.7	554.5	599.5	623.8
	3.6	772.4	835.6	839.3	914.1	955.5
	4.8	918.3	1001	1008	1103	1156
	6.0	964.2	1054	1062	1164	1222
1.0	1.2	86.3	91.3	91.4	98.1	102
	2.4	204.2	221.1	221.2	242.9	255.2
	3.6	274.1	302.4	303.3	339.4	359.9
	4.8	301.9	337.5	340.1	384.3	410.2
	6.0	308.3	346.3	349.2	396.4	424.1
2.0	1.2	36.6	38.1	38.2	41.4	43.0
	2.4	73.7	79.2	79.7	88.8	94.1
	3.6	84.5	96.1	96.1	109	116
	4.8	86.7	97.4	97.8	111	119
	6.0	87.3	98.1	98.3	113	121

are maximum and Table 3.5 shows how their magnitudes vary depending on the change in parameters of orthotropy and shallowness degree.

From this table we can see that the magnitudes of the stresses $\sigma_{\bar{b},l}$ for variant 5 at $f = 0, 0.5, 1, 2$ are 1.07, 1.29, 1.41, 1.44 times that in magnitude of variant 1. If for variant 1 the ratio $\sigma_{\bar{b},l}/\sigma_{\bar{b},m}$ for $f = 0, 0.5, 1, 2$ takes the values 1.35, 1.03, 0.91, and 0.84, then for variant 5 this ratio is equal to 5.08, 4.13, 3.69, and 3.41, respectively. The magnitudes of these ratios show how variations of the parameters of orthotropy and of the degree of shallowness will simultaneously affect the stress state of a laminated shell.

3.2 Noncircular Cylindrical Shells

3.2.1 Governing Equations

Together with circular shells noncircular ones (Fig. 3.2) have come into widespread use in various branches of engineering, for example, in aircraft building,

Table 3.5 The magnitudes of the stresses

f	N	σ_y/q_0 $(x = 6, y = 5)$					
		$\sigma^+_{u3,6}$	$\sigma^-_{u3,6}$	σ^+_{op}	σ^-_{op}	$\sigma^+_{u3,H}$	$\sigma^-_{u3,H}$
0	1	−233	−93.3	−172	172	93.3	233
	2	−240	−96.0	−120	120	96.0	240
	3	−241	−96.5	−96.5	96.5	96.5	241
	4	−246	−98.4	−68.0	68.0	98.4	246
	5	−249	−100	−45.8	45.8	100	249
0.5	1	−60.3	−10.6	−19.7	102.3	55.6	105
	2	−66.7	−11.9	−15.0	76.6	60.1	116
	3	−68.8	−12.8	−12.8	61.8	61.8	118
	4	−75.0	−13.8	−9.5	46.6	67.8	129
	5	−79.2	−14.8	−6.8	32.7	71.2	135
1.0	1	−15.8	2.7	5.0	50.5	27.5	46.0
	2	−18.2	2.8	3.7	39.0	31.1	52.2
	3	−19.2	2.9	2.2	31.8	31.8	53.6
	4	−21.6	3.0	2.1	24.6	35.8	60.4
	5	−23.5	3.6	1.4	17.6	38.3	64.9
2.0	1	−4.0	3.7	6.8	25.8	14.0	21.8
	2	−4.6	4.2	5.3	20.0	16.0	24.8
	3	−4.7	4.4	4.4	16.5	16.5	25.6
	4	−5.5	4.8	3.3	12.8	18.7	29.0
	5	−6.2	5.1	2.3	9.2	20.1	31.4

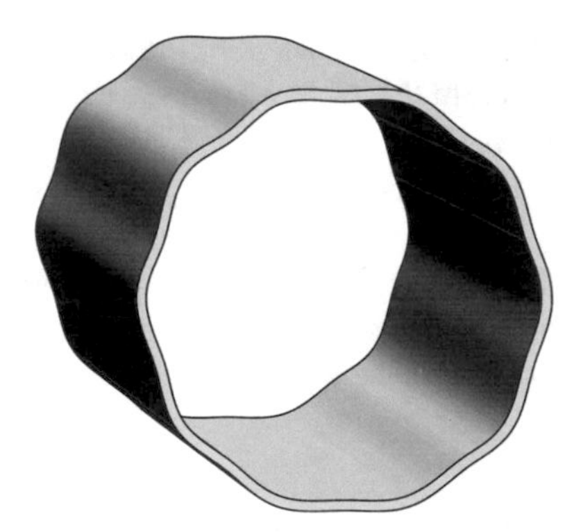

Fig. 3.2 Noncircular shell

shipbuilding, space technology, oil and gas industry, building and structural designs. Investigation of the stress-strain state of noncircular cylindrical shells of variable thickness is of theoretical as well as of applied interest.

By using the Donnell-Vlasov equations (Donell [2], Vlasov [20]) we investigate the stress-strain state of noncircular isotropic thin cylindrical shells with varying thickness along a generatrix and directrix under arbitrary surface loads (Soldatos [18]). The shells may be closed or open along the directrix. The relations that describe the deformation for that class of shells have the following form (Grigorenko and Zakhariichenko [10]) in the coordinate system *s*, *t*, where *s* and *t* are the arc lengths along the generatrix and directrix; the expressions for the strains:

$$\begin{aligned} &\varepsilon_s = \frac{\partial u}{\partial s}, \quad \varepsilon_t = \frac{\partial v}{\partial t} + \frac{w}{R}, \quad \varepsilon_{st} = \frac{\partial u}{\partial t} + \frac{\partial v}{\partial s}, \\ &\kappa_s = -\frac{\partial^2 w}{\partial s^2}, \quad \kappa_t = -\frac{\partial^2 w}{\partial t^2}, \quad \kappa_{st} = -\frac{\partial^2 w}{\partial s \partial t}; \end{aligned} \tag{3.2.1}$$

the equilibrium conditions:

$$\begin{aligned} &\frac{\partial N_s}{\partial s} + \frac{\partial S}{\partial t} = 0, \quad \frac{\partial S}{\partial t} + \frac{\partial N_t}{\partial t} = 0, \\ &\frac{\partial Q_s}{\partial s} + \frac{\partial Q_t}{\partial t} - \frac{1}{R} N_t + q_\gamma = 0, \\ &\frac{\partial M_s}{\partial s} + \frac{\partial H}{\partial t} - Q_s = 0, \quad \frac{\partial M_t}{\partial t} + \frac{\partial H}{\partial s} - Q_t = 0; \end{aligned} \tag{3.2.2}$$

the elasticity relations:

$$\begin{aligned} &N_s = D_N(\varepsilon_s + \nu\varepsilon_t), \quad N_t = D_N(\varepsilon_t + \nu\varepsilon_s), \quad S = \frac{1-\nu}{2} D_N \varepsilon_{st}, \\ &M_s = D_M(\kappa_s + \nu\kappa_t), \quad M_t = D_M(\kappa_t + \nu\kappa_s), \quad H = (1-\nu) D_M \kappa_{st}, \end{aligned} \tag{3.2.3}$$

where $R = R(t)$ is the radius of the directix curvature, $D_N = \frac{Eh(s,t)}{1-\nu^2}$ and $D_M = \frac{Eh^3(s,t)}{12(1-\nu^2)}$ are the stiffnesses.

In Eqs. (3.2.1)–(3.2.3) u, v, w are the displacements along the generatrix, directrix, and normal to a median surface, $\varepsilon_s, \varepsilon_t, \varepsilon_{st}, \kappa_s, \kappa_t, \kappa_{st}$ are the tangential and bending strains, N_s, N_t, S, Q_s, Q_t are the forces, M_s, M_t, H are the moments, $h = h(s, t)$ is the shell thickness, E and ν are the elastic modulus and Poisson's ratio, $q_\gamma = q_\gamma(s,t)$ is the surface load.

After some rearrangements, we obtain from the expressions (3.2.1)–(3.2.3) the governing system of equations for displacements:

$$
\begin{aligned}
& D_N\left\{\frac{\partial}{\partial s}\left[\frac{\partial u}{\partial s}+\nu\left(\frac{\partial u}{\partial s}+\frac{w}{R}\right)\right]+\frac{1-\nu}{2}\frac{\partial}{\partial t}\left(\frac{\partial u}{\partial t}+\frac{\partial v}{\partial s}\right)\right\} \\
& \quad+\frac{\partial D_N}{\partial s}\left[\frac{\partial u}{\partial s}+\nu\left(\frac{\partial v}{\partial t}+\frac{w}{R}\right)\right]+\frac{1-\nu}{2}\frac{\partial D_N}{\partial t}\left(\frac{\partial u}{\partial t}+\frac{\partial v}{\partial s}\right)=0, \\
& D_N\left[\frac{\partial}{\partial t}\left(\frac{\partial v}{\partial t}+\frac{w}{R}+\nu\frac{\partial u}{\partial s}\right)+\frac{1-\nu}{2}\frac{\partial}{\partial s}\left(\frac{\partial u}{\partial t}+\frac{\partial v}{\partial s}\right)\right] \\
& \quad+\frac{\partial D_N}{\partial t}\left(\frac{\partial v}{\partial t}+\frac{w}{R}+\nu\frac{\partial u}{\partial s}\right)+\frac{1-\nu}{2}\frac{\partial D_N}{\partial s}\left(\frac{\partial u}{\partial t}+\frac{\partial v}{\partial s}\right)=0, \\
& \frac{\partial}{\partial s}\left[D_M\frac{\partial \Delta w}{\partial s}+\frac{\partial D_M}{\partial t}(1-\nu)\frac{\partial^2 w}{\partial s\partial t}+\frac{\partial D_M}{\partial s}\left(\frac{\partial^2 w}{\partial s^2}+\nu\frac{\partial^2 w}{\partial t^2}\right)\right] \\
& \quad+\frac{\partial}{\partial t}\left[D_M\frac{\partial \Delta w}{\partial t}+\frac{\partial D_M}{\partial s}(1-\nu)\frac{\partial^2 w}{\partial s\partial t}+\frac{\partial D_M}{\partial t}\left(\frac{\partial^2 w}{\partial t^2}+\nu\frac{\partial^2 w}{\partial s^2}\right)\right] \\
& \quad+\frac{D_N}{R}\left(\frac{\partial v}{\partial t}+\frac{w}{R}+\nu\frac{\partial u}{\partial s}\right)=q_\gamma, \quad 0\le s\le L, \quad 0\le t\le 2\pi.
\end{aligned}
\tag{3.2.4}
$$

In the case of shells that are closed along the directrix, boundary conditions are specified on the curvilinear edges, and for open shells on curvilinear and rectilinear edges. The boundary conditions can be formulated in terms of displacements or in mixed form. Arbitrary boundary conditions are specified on the curvilinear edges. In the case of open cylindrical shells the following boundary conditions on the rectilinear edges are considered: either rigid or hinged fixation of both edges. These conditions are:

$$u=v=w=\vartheta_t=0 \text{ at } t=t_1, \quad t=t_2, \tag{3.2.5}$$

or:

$$u=v=w=M_t=0 \text{ at } t=t_1, \quad t=t_2, \tag{3.2.6}$$

i.e., either the fixation of one edge is a hinge, Eq. (3.2.6), or the other edge is rigid, Eq. (3.2.5). In general, other boundary conditions on the rectilinear edges can also be specified.

After some rearrangements the resulting system of differential equations (3.2.4) can be written as follows:

$$
\begin{aligned}
\frac{\partial^2 u}{\partial s^2} &= a_{11}\frac{\partial^2 u}{\partial t^2}+a_{12}\frac{\partial u}{\partial t}+a_{13}\frac{\partial u}{\partial s}+a_{14}\frac{\partial v}{\partial t}+a_{15}\frac{\partial^2 v}{\partial s\partial t}+a_{16}\frac{\partial v}{\partial s}+a_{17}w+a_{18}\frac{\partial w}{\partial s}, \\
\frac{\partial^2 v}{\partial s^2} &= a_{21}\frac{\partial u}{\partial t}+a_{22}\frac{\partial u}{\partial s}+a_{23}\frac{\partial^2 v}{\partial s\partial t}+a_{24}\frac{\partial^2 v}{\partial t^2}+a_{25}\frac{\partial v}{\partial s}+a_{26}\frac{\partial v}{\partial s}+a_{27}w+a_{28}\frac{\partial w}{\partial t}, \\
\frac{\partial^4 w}{\partial s^4} &= a_{31}\frac{\partial u}{\partial s}+a_{32}\frac{\partial v}{\partial t}+a_{33}w+a_{34}\frac{\partial^2 w}{\partial t^2}+a_{35}\frac{\partial^3 w}{\partial t^3}+a_{36}\frac{\partial^3 w}{\partial t^3}+a_{37}\frac{\partial^4 w}{\partial t^4} \\
&\quad+a_{38}\frac{\partial^2 w}{\partial s\partial t}+a_{39}\frac{\partial^3 w}{\partial s\partial t}+a_{3,10}\frac{\partial^4 w}{\partial s^2\partial t}+a_{3,11}\frac{\partial^3 w}{\partial s^2\partial t}+a_{3,12}\frac{\partial^2 w}{\partial s^2}+a_{3,13}\frac{\partial^3 w}{\partial s^3}+\frac{q_\gamma}{D_M},
\end{aligned}
\tag{3.2.7}
$$

with the coefficients $a_{ij}, i = 1,2,3; j = 1,2,\ldots,13$ being expressed in terms of mechanical characteristics that are functions of the coordinates s and t.

We will seek the solution of the boundary-value problem for the system of partial differential equations (3.2.7) with appropriate boundary conditions on the boundary in the form:

$$u(s,t) = \sum_{i=0}^{N} u_i(s)\phi_i(t), \quad v(s,t) = \sum_{i=0}^{N} v_i(s)\phi_i(t), \quad w(s,t) = \sum_{i=0}^{N} w_i(s)\psi_i(t), \tag{3.2.8}$$

where $u_i(s), v_i(s), w_i(s), i = \overline{0,N}$ are the unknown functions, $\phi_i(t)$ and $\psi_i(t)$ are linear combinations of B-splines of the third and fifth power, respectively, which satisfy the boundary conditions on the rectilinear contours for open shells and periodicity conditions for closed shells.

Expressions for B-splines of the third and fifth power at the augmented mesh, if numerated in the middle node of a carrier, have the form:

$$B_3^i(t) = \frac{1}{6}\begin{cases} 0, & t < t_{i-2}, \\ z^3, & t_{i-2} \le t < t_{i-1}, \\ 1 + 3z + 3z^2(1-z), & t_{i-1} \le t < t_i, \\ 3z^3 - 6z^2 + 4, & t_i \le t < t_{i+1}, \\ (1-z)^3, & t_{i+1} \le t < t_{i+2}, \\ 0, & t \ge t_{i+2}, \end{cases} \tag{3.2.9}$$

and:

$$B_5^i(t) = \frac{1}{120}\begin{cases} 0, & t < t_{i-3}, \\ z^5, & t_{i-3} \le t < t_{i-2}, \\ -5z^5 + 5z^4 - 10z^3 - 10z^2 + 5z + 1, & t_{i-2} \le t < t_{i-1}, \\ 10z^5 - 20z^4 - 20z^3 + 20z^2 + 50z + 26, & t_{i-1} \le t < t_i, \\ -10z^5 + 30z^4 - 60z^2 + 66, & t_i \le t < t_{i+1}, \\ 5z^5 - 20z^4 + 20z^3 + 20z^2 - 50z + 26, & t_{i+1} \le t < t_{i+2}, \\ (1-z)^5, & t_{i+2} \le t < t_{i+3}, \\ 0, & t \ge t_{i+3}, \end{cases} \tag{3.2.10}$$

where $z = (t - t_k)/h \ (t_{k+1} - t_k = h = \mathrm{const})$ on the interval $[t_k, t_{k+1}]$,

$k = \overline{i - (m+1)/2, i + (m+1)/2 - 1}$, $i = \overline{-(m+1)/2 + 1, N + (m+1)/2 - 1}, m = 3; 5.$

In particular, if the edge $t = t_1$ is clamped and the edge $t = t_2$ is hinged, the following relations hold:

$$
\begin{aligned}
\phi_0(t) &= -4B_3^{-1}(t) + B_3^0(t), \\
\phi_1(t) &= B_3^{-1}(t) - \frac{1}{2}B_3^0(t) + B_3^1(t), \\
\phi_i(t) &= B_3^i(t), \quad i = \overline{2, N-2}, \\
\phi_{N-1}(t) &= B_3^{N-1}(t) - \frac{1}{2}B_3^N(t) + B_3^{N+1}(t), \\
\phi_N(t) &= B_3^N(t) - 4B_3^{N+1}(t), \\
\psi_0(t) &= \frac{165}{4}B_5^{-2} - \frac{33}{8}B_5^{-1}(t) + B_5^0(t), \\
\psi_1(t) &= B_5^{-1}(t) - \frac{26}{33}B_5^0(t) + B_5^1(t), \\
\psi_2(t) &= B_5^{-2}(t) - \frac{1}{33}B_5^0(t) + B_5^2(t), \\
\psi_i(t) &= B_5^i(t), \quad i = \overline{3, N-3}, \\
\psi_{N-2}(t) &= B_5^{N-2}(t) - B_5^{N+2}(t), \quad \psi_{N-1}(t) = B_5^{N-1}(t) - B_5^{N+1}(t), \\
\psi_N(t) &= B_5^N(t) - 3B_5^{N+1}(t) + 12B_5^{N+2}(t).
\end{aligned}
\tag{3.2.11}
$$

From (3.2.11) it is easy to derive expressions for $\phi_i(t)$ and $\psi_i(t)$ when both edges are clamped or hinged. Expressions in the form of linear combinations of B-splines under other boundary or symmetry conditions are constructed similarly.

Having chosen the functions $\phi_i(t)$ and $\psi_i(t)$ in such a way that the boundary or symmetry conditions are satisfied, we substitute the expressions (3.2.8) into the differential equations (3.2.7) and require them to be satisfied at the collocation points $t = t_k$, $k = \overline{0, N}$. After some rearrangements, we arrive at the system of ordinary differential equations of the $8(N + 1)$th order:

$$
\frac{d\overline{Z}}{ds} = A(s)\overline{Z} + \overline{f}(s), \quad 0 \leq s \leq L, \tag{3.2.12}
$$

where:

$$
\begin{aligned}
\overline{Z} &= \{\bar{z}_1, \bar{z}_2, \ldots, \bar{z}_8\}^{\mathrm{T}} = \{\bar{u}, \bar{u}', \bar{v}, \bar{v}', \bar{w}, \bar{w}', \bar{w}'', \bar{w}'''\}^{\mathrm{T}}, \\
\bar{z}_m &= \{z_{m_0}, z_{m_1}, \ldots, z_{m_N}\}^{\mathrm{T}}, \quad m = \overline{1, 8}.
\end{aligned}
$$

Starting from the boundary conditions specified on the edges $s = 0$ and $s = L$, we can formulate boundary conditions for the system of equations (3.2.12), which in the general case become:

$$
B_1\overline{Z}(0) = \bar{b}_1, \quad B_2\overline{Z}(L) = \bar{b}_2. \tag{3.2.13}
$$

In order to solve the boundary-value problem for the system of equations (3.2.12) with the boundary conditions (3.2.13), we will use the stable method of discrete orthogonalization. By substitution of the found values of functions

$u_i(s), v_i(s)$ and $w_i(s), i = \overline{0, N}$ into the expressions (3.2.8), we obtain the solution of the initial problem for the displacements, from which all the parameters of the shell stress-strain state can be calculated.

We now turn to the analysis of cylindrical shells with elliptical cross-section. Specifically, we will consider the stress-strain problem of an isotropic cylindrical shell with an elliptical cross-section under a uniformly distributed normal surface load q_γ (Ding and Tang [1], Grigorenko et al. [7]). We specify the parametric equations of the shell cross-section by $x = b\cos\psi, z = a\sin\psi (0 \le \psi \le 2\pi)$ where a and b are the minor and major semi-axes of the ellipse. We will restrict our analysis to shells whose cross-sectional perimeter of the median surface is kept unchanged and equal to the perimeter of the circle with radius R, i.e., $\pi(a+b)f(\Delta) = 2\pi R$, where:

$$\Delta = \frac{b-a}{b+a}, \quad a = \frac{R}{f}(1-\Delta), \quad b = \frac{R}{f}(1+\Delta),$$
$$f(\Delta) = 1 + \frac{\Delta^2}{4} + \frac{\Delta^4}{64} + \frac{\Delta^6}{256} + \cdots, \quad \frac{a}{b} = \frac{1-\Delta}{1+\Delta}, \tag{3.2.14}$$
$$\frac{\mathrm{d}}{\mathrm{d}t} = \frac{1}{\gamma(\psi)}\frac{\mathrm{d}}{\mathrm{d}\psi}, \quad \gamma(\psi) = \sqrt{\left(\frac{\mathrm{d}x}{\mathrm{d}\psi}\right)^2 + \left(\frac{\mathrm{d}z}{\mathrm{d}\psi}\right)^2}.$$

We will investigate three variants of the stress-strain state of an elliptic cylindrical shell:

1. the shell is open $(-\frac{\pi}{2} \le \psi \le \frac{\pi}{2})$, its curvilinear and rectilinear edges are rigidly fixed, i.e., the boundary conditions
$$u = v = w = \frac{\partial w}{\partial s} = 0 \quad \text{at} \quad s = 0, \quad s = L,$$
$$u = v = w = \frac{\partial w}{\partial t} = 0 \quad \text{at} \quad \psi = -\frac{\pi}{2}, \quad \psi = \frac{\pi}{2} \tag{3.1.15}$$
atare relevant;
2. the shell is open $(-\frac{\pi}{2} \le \psi \le \frac{\pi}{2})$, its curvilinear edges are rigidly fixed and the rectilinear ones are hinged, i.e., the boundary conditions
$$u = v = w = \frac{\partial w}{\partial s} = 0 \text{ at } s = 0, \quad s = L,$$
$$u = v = w = M_t = 0 \text{ at } \psi = -\frac{\pi}{2}, \quad \psi = \frac{\pi}{2} \tag{3.2.16}$$
hold;
3. the shell is closed, its curvilinear edges are rigidly fixed, i.e., the boundary conditions are:
$$u = v = w = \frac{\partial w}{\partial s} = 0 \text{ at } s = 0, \quad s = L. \tag{3.2.17}$$

In solving the problem, we may consider the part of the shell, $0 \leq \psi \leq \pi/2$, by specifying symmetry conditions at $s = L/2, \psi = 0$ and $\psi = \pi/2$.

For solving the problem we used the following data: $R = 20$, $L = 60$, $q_\gamma = q_0 =$ const, $h = h_0(1 + \alpha|\sin\psi|), h_0 = 0.5, v = 0.3$; the values of Δ and α are given in Tables 3.6 and 3.7.

Table 3.6 summarizes the maximum values of the deflection w in the section $s = L/2$ for the variants (1) and (2) at $\psi = \frac{0.7\pi}{2}$ and for variant 3 at $\psi = \frac{\pi}{2}$, of the force N_t for variants (1) and (2) at $\psi = 0$, and for variant (3) at $\psi = \frac{\pi}{2}$; of the moment M_t for variants (1) and (3) at $\psi = \frac{\pi}{2}$ and for variant (2) at $\psi = \frac{0.8\pi}{2}$.

From the data presented in the table, it can be seen how the deflection, force, and moment for the three variants under different values of Δ vary depending on the variation in the shell thickness along the directrix.

Table 3.7 presents data concerning distribution of the deflection, force, and moment along the directrix for the closed shell at $s = L/2$ with the same parameters, as in the previous case, for different values of α at $\Delta = 0.005$.

We now pose the problem to provide the most uniform distribution of deflection w along the directrix by choosing adequate parameters of the variable thickness.

As it follows from the table, such a state is reached for $\alpha = 0.83$. Also, we can see how values of the force N_t and of the moment M_t redistribute along the directrix depending on the variation of the thickness.

Table 3.6 Maximum values of the deflection w in the section $s = L/2$

Δ	n	$\alpha =$	$\alpha = 0.1$	$\alpha = 0.25$	$\alpha = 0.5$
$Ew/10^4 q_0$					
0.1	1	0.4440	0.3985	0.3484	0.2891
	2	0.4600	0.4213	0.3723	0.3084
	3	0.5363	0.4845	0.4233	0.3498
0.2	1	0.9409	0.8441	0.7227	0.5683
	2	0.9673	0.8894	0.7875	0.6511
	3	1.246	1.120	0.9718	0.7956
$N_t/10^3 q_0$					
0.1	1	2.256	2.169	2.032	1.806
	2	2.121	2.062	1.967	1.805
	3	1.770	1.755	−1.738	−1.719
0.2	1	4.418	4.294	4.072	3.659
	2	4.183	4.108	3.965	3.677
	3	−3.661	−3.609	−3.547	−3.470
$M_t/10^3 q_0$					
0.1	1	−1.633	−1.858	−2.205	−2.801
	2	0.6614	0.7377	0.8548	1.056
	3	0.2542	0.2783	0.3168	0.3882
0.2	1	−3.913	4.391	−5.127	−6.382
	2	1.690	1.856	2.130	2.583
	3	0.7512	0.8084	0.8998	1.070

Table 3.7 Data concerning distribution of the deflection, force, and moment along the directrix

$\psi/\frac{\pi}{2}$	$\Delta=0\ \alpha=0$	$\Delta=0.005\ \alpha=0$	$\Delta=0.005\ \alpha=0.25$	$\Delta=0.005\ s=s_0$	$\Delta=0.005\ \alpha=0.83$
s_N					
0	0.7334	0.5492	0.5322	0.5150	0.4934
0.2	0.7334	0.5841	0.5548	0.5281	0.4968
0.4	0.7334	0.6761	0.6087	0.5551	0.4988
0.6	0.7334	0.7911	0.6694	0.5812	0.4961
0.8	0.7334	0.8853	0.7155	0.5995	0.4929
1.0	0.7334	0.9215	0.7327	0.6061	0.4918
$N_t/10^2q_0$					
0	0.2000	1.081	1.170	1.261	1.374
0.2	0.2000	0.9129	0.8439	0.7825	0.7086
0.4	0.2000	0.4723	0.3590	0.2555	0.1330
0.6	0.2000	−0.0722	−0.1406	−0.1988	−0.2634
0.8	0.2000	−0.5133	−0.5119	−0.5029	−0.4860
1.0	0.2000	−0.6820	−0.6493	−0.6106	−0.5582
$\Delta=0.025\sqrt{\pi}$					
0	0	−0.8367	−0.5589	−0.3460	−0.1224
0.2	0	−0.6832	−0.4657	−0.2431	0.0601
0.4	0	−0.2765	−0.1084	0.0456	0.2013
0.6	0	0.2487	0.3026	0.2566	0.0101
0.8	0	0.6939	0.6537	0.4396	−0.1447
1.0	0	0.8710	0.8012	0.5320	−0.1640

Thus, the obtained results show the influence of thickness variation in noncircular cylindrical shells on their stress-strain state.

We will now consider stress-strain problems for cylindrical shells of variable thickness with corrugated cross-section under a surface load (Grigorenko et al. [8]). The corresponding equations have the form:

$$\begin{aligned}
\frac{\partial^2 u}{\partial\varphi^2} &= a_{11}\frac{\partial^2 u}{\partial\partial s^2}+a_{12}\frac{\partial u}{\partial s}+a_{13}\frac{\partial u}{\partial\varphi}+a_{14}\frac{\partial v}{\partial s}+a_{15}\frac{\partial^2 v}{\partial s\partial\varphi}+a_{16}\frac{\partial v}{\partial\varphi}+a_{17}w+a_{18}\frac{\partial w}{\partial s},\\
\frac{\partial^2 v}{\partial\varphi^2} &= a_{21}\frac{\partial u}{\partial s}+a_{22}\frac{\partial u}{\partial\varphi}+a_{23}\frac{\partial^2 u}{\partial s\partial\varphi}+a_{24}\frac{\partial^2 v}{\partial s^2}+a_{25}\frac{\partial v}{\partial s}+a_{26}\frac{\partial v}{\partial\varphi}+a_{27}w+a_{28}\frac{\partial w}{\partial\varphi},\\
\frac{\partial^4 w}{\partial\varphi^4} &= a_{31}\frac{\partial u}{\partial s}+a_{32}\frac{\partial v}{\partial\varphi}+a_{33}w+a_{34}\frac{\partial^2 w}{\partial s^2}+a_{35}\frac{\partial^3 w}{\partial s^3}+a_{36}\frac{\partial^4 w}{\partial s^4}+a_{37}\frac{\partial^2 w}{\partial s\partial\varphi}\\
&\quad+a_{38}\frac{\partial^3 w}{\partial s^2\partial\varphi}+a_{39}\frac{\partial w}{\partial\varphi}+a_{3,10}\frac{\partial^3 w}{\partial s\partial\varphi^2}+a_{3,11}\frac{\partial^4 w}{\partial s^2\partial\varphi^2}+a_{3,12}\frac{\partial^2 w}{\partial\varphi^2}\\
&\quad+a_{3,13}\frac{\partial^3 w}{\partial\varphi^3}+b_1,\quad 0\le s\le L,\quad 0\le\phi\le 2\pi,
\end{aligned} \tag{3.2.18}$$

where:

$$
\begin{aligned}
a_{11} &= -\frac{2}{\gamma_{22}(1-\nu)}, \quad a_{12} = -\frac{2}{D_N\gamma_{22}(1-\nu)}\frac{\partial D_N}{\partial s},\\
a_{13} &= -\frac{1}{D_N\gamma_{22}}\left(\gamma_{11}^2\frac{\partial D_N}{\partial\phi}+\gamma_{21}D_N\right), \quad a_{14} = -\frac{\gamma_{11}}{D_N\gamma_{22}}\frac{\partial D_N}{\partial\phi}, \quad a_{15} = -\frac{(1+\nu)\gamma_{11}}{(1-\nu)\gamma_{22}},\\
a_{16} &= -\frac{2\nu\gamma_{11}}{D_N\gamma_{22}(1-\nu)}\frac{\partial D_N}{\partial s}, \quad a_{17} = -\frac{2\nu}{D_N R_\phi\gamma_{22}(1-\nu)}\frac{\partial D_N}{\partial s},\\
a_{18} &= -\frac{2\nu}{D_N R_\phi\gamma_{22}(1-\nu)}, \quad a_{21} = -\frac{\nu\gamma_{11}}{\gamma_{22}D_N}\frac{\partial D_N}{\partial\phi}, \quad a_{22} = -\frac{(1-\nu)\gamma_{11}}{2D_N\gamma_{22}}\frac{\partial D_N}{\partial s},\\
a_{23} &= -\frac{(1+\nu)\gamma_{11}}{2\gamma_{22}}, \quad a_{24} = -\frac{1-\nu}{2\gamma_{22}}, \quad a_{25} = -\frac{1-\nu}{2\gamma_{22}D_N}\frac{\partial D_N}{\partial s},\\
a_{26} &= -\frac{1}{D_N\gamma_{22}}\left(\gamma_{11}^2\frac{\partial D_N}{\partial\phi}+\gamma_{21}D_N\right), \quad a_{27} = -\frac{\gamma_{11}}{R_\phi D_N\gamma_{22}}\left(\frac{\partial D_N}{\partial\phi}-\frac{D_N}{R_\phi}\frac{\partial R_\phi}{\partial\phi}\right),\\
a_{28} &= -\frac{\gamma_{11}}{R_\phi\gamma_{22}}, \quad a_{31} = -\frac{\nu D_N}{D_M R_\phi\gamma_{44}}, \quad a_{32} = -\frac{D_N\gamma_{11}}{D_M R_\phi\gamma_{44}}, \quad a_{33} = -\frac{D_N}{D_M R_\phi^2\gamma_{44}},\\
a_{34} &= -\frac{1}{D_M\gamma_{44}}\left(\frac{\partial^2 D_M}{\partial s^2}+\nu\gamma_{21}\frac{\partial D_M}{\partial\phi}+\nu\gamma_{22}\frac{\partial^2 D_M}{\partial\phi^2}\right), \quad a_{35} = -\frac{2}{D_M\gamma_{44}}\frac{\partial D_M}{\partial s},\\
a_{36} &= -\frac{1}{\gamma_{44}}, \quad a_{37} = -\frac{2}{D_M\gamma_{44}}\left(\gamma_{11}^2(1-\nu)\frac{\partial^2 D_M}{\partial s\partial\phi}+\gamma_{21}\frac{\partial D_M}{\partial s}\right),\\
a_{38} &= -\frac{2}{D_M\gamma_{44}}\left(\gamma_{11}^2\frac{\partial D_M}{\partial\phi}+\gamma_{21}D_M\right),\\
a_{39} &= -\frac{1}{D_M\gamma_{44}}\times\left[\frac{\partial D_M}{\partial\varphi}\left(\gamma_{21}^2+2\gamma_{11}\gamma_{31}\right)+\gamma_{21}\gamma_{22}\frac{\partial^2 D_M}{\partial\varphi^2}+\nu\gamma_{21}\frac{\partial^2 D_M}{\partial s^2}+\gamma_{41}D_M\right],\\
a_{3,10} &= -\frac{2\gamma_{22}}{D_M\gamma_{44}}\frac{\partial D_M}{\partial s}, \quad a_{3,11} = -\frac{2\gamma_{22}}{\gamma_{44}},\\
a_{3,12} &= -\frac{1}{D_M\gamma_{44}}\times\left[\frac{\partial D_M}{\partial\varphi}\left(\gamma_{21}\gamma_{22}+2\gamma_{11}\gamma_{32}\right)+\gamma_{22}^2\frac{\partial^2 D_M}{\partial\varphi^2}+\nu\gamma_{22}\frac{\partial^2 D_M}{\partial s^2}+\gamma_{42}D_M\right],\\
a_{3,13} &= -\frac{1}{D_M\gamma_{44}}\left(2\gamma_{11}\gamma_{33}\frac{\partial D_M}{\partial\phi}+\gamma_{43}D_M\right), \quad b_1 = \frac{q_\gamma}{D_M\gamma_{44}},\\
\gamma &= \gamma(\phi) = \sqrt{r^2+(r')^2}, \quad R_\phi = \frac{\left[r^2+(r')^2\right]^{\frac{3}{2}}}{r^2+2(r')^2-rr''},\\
\gamma_{11} &= \frac{1}{\gamma}, \quad \gamma_{21} = -\frac{\gamma'}{\gamma^3}, \quad \gamma_{22} = \frac{1}{\gamma^2}, \quad \gamma_{31} = \frac{3(\gamma')^2}{\gamma^5}-\frac{\gamma''}{\gamma^4}, \quad \gamma_{32} = -\frac{3\gamma'}{\gamma^4}, \quad \gamma_{33} = \frac{1}{\gamma^3},\\
\gamma_{41} &= \frac{10\gamma'\gamma''}{\gamma^6}-\frac{15(\gamma')^3}{\gamma^5}-\frac{\gamma'''}{\gamma^5}, \quad \gamma_{42} = \frac{15(\gamma')^2}{\gamma^6}-\frac{4\gamma''}{\gamma^5}, \quad \gamma_{43} = -\frac{6\gamma'}{\gamma^5}, \quad \gamma_{44} = \gamma^{-4}.
\end{aligned}
\tag{3.2.19}
$$

In Eq. (3.2.18) a polar coordinate system $r(\varphi)$ was adopted for the shell cross-section, where r denotes the polar radius, and φ is the polar angle. In addition the following designations were used: s is the longitudinal coordinate, u, v, w are the displacements along the generatrix, directrix, and normal, respectively, $D_N = Eh/(1-\nu^2)$ and $D_M = Eh^3/[12(1-\nu^2)]$ are the tangential and bending stiffnesses, $h(s,\phi)$ is the shell thickness, E and ν are Young's modulus and Poisson's ratio, and qγ is the transverse load.

Boundary conditions for displacements are specified on the shell ends $s=0$ and $s=L$. We denote the polar radius and shell thickness as follows:

$$r = r_0 + \alpha\cos k\phi \quad (r_0 = \text{const.}), \quad h(\phi) = h_0(1-\beta\cos k\phi) \\ (h_0 = \text{const.}, 0 \le \phi \le 2\pi), \tag{3.2.20}$$

where r_0 is the radius of a circle, α is the amplitude, k is the corrugation frequency, β is the factor of variability of the thickness along the directrix. We assume that the shell is rigidly fixed at the ends, i.e., the conditions

$$u = v = w = \vartheta_s = 0 \tag{3.2.21}$$

are met at $s=0$ and $s=L$.

We will seek the solution of the boundary-value problem for the system of partial differential equations (3.2.18) in the form:

$$\begin{aligned} u(s,\phi) &= \sum_{i=0}^{N} u_i(\phi)\psi_{1i}(s), \\ v(s,\phi) &= \sum_{i=0}^{N} v_i(\phi)\psi_{2i}(s), \\ w(s,\phi) &= \sum_{i=0}^{N} w_i(\phi)\psi_{3i}(s), \end{aligned} \tag{3.2.22}$$

where $u_i(\varphi)$, $v_i(\varphi)$, $w_i(\varphi)$, $i=\overline{0,N}$ are the unknown functions depending on the coordinate φ in the shell cross-section, ψ_{ni}, $n = 1,,2,3$ are linear combinations of third-power B-splines for $n=1,2$ and of the fifth power for $n=3$, with the help of which we can satisfy the boundary conditions on ends accurately [3, 7]. By substituting the expressions (3.2.22) into the differential equations of the system (3.2.18) and by requiring them to be satisfied at collocation points $s=s_k$ $(k=\overline{0,N})$, i.e., in the section $N+1$, we obtain the following system of ordinary differential equations after some transformations:

$$\frac{d\bar{U}}{d\phi} = A(\phi)\bar{U} + \bar{f}(\phi), \quad 0 \le \phi \le 2\pi, \tag{3.2.23}$$

where:

$$\bar{U} = \{\bar{u}, \bar{u}', \bar{v}, \bar{v}', \bar{w}, \bar{w}', \bar{w}'', \bar{w}'''\}^{\mathrm{T}}$$
$$= \{\bar{u}_1, \bar{u}_2, \bar{u}_3, \bar{u}_4, \bar{u}_5, \bar{u}_6, \bar{u}_7, \bar{u}_8\}^{\mathrm{T}},$$
$$\bar{u}_m = \{u_{m_0}, \ldots, u_{m_N}\}^{\mathrm{T}}, m = \overline{1, 8}.$$

In this case, the periodicity or symmetry conditions are formulated for the closed shell along the directrix. For the open shell the boundary conditions are specified on the rectilinear edges.

We will solve the boundary-value problem for the system of equations (3.2.23) with associated conditions in φ by the stable numerical method of discrete orthogonalization.

After substituting the determined values of functions $u_i(\varphi)$, $v_i(\varphi)$, $w_i(\varphi)$, $i = \overline{0, N}$ into the expressions (3.2.22), we obtain the solution of the initial problem for the displacements. Based on this solution, we calculate all parameters of the stress-strain state of the shell.

3.2.2 Results

We now consider how a variation in the corrugation amplitude under fixed corrugation frequency affects the stress-strain state of cylindrical shells. We assume that the cylindrical constant-thickness shell is rigidly fixed at the ends and that the initial data have the following values:

$$r_0 = 15, L = 30, \quad r_0 = 0.5, \quad \nu = 0.3,$$
$$q_\gamma = q_0 = \text{const.}, \quad k = 4, 8, \quad \alpha = 0; 0.1; 0.3; 0.5; \quad 0 \leq \varphi \leq \pi/4$$

Figures 3.3 and 3.4 show the plots of the distribution of the deflection w and of the bending moment M_φ along the directrix for $k = 4$ (solid lines) and for $k = 8$ (dashed lines) in the section $L/2$. From Fig. 3.3 it can be seen that with increasing parameter α at $k = 4$, the deflection of the apex of the corrugation convex part ($\varphi = 0$) rises by a factor of 5.7 and 9.9 at $\alpha = 0.3$ and $\alpha = 0.5$, respectively, in comparison with the deflection at $\alpha = 0.1$, where this part of the corrugation deflects in opposition to the applied load. The concave part of the corrugation ($\varphi = \pi/4$) displaces in the direction of the applied load and, as result, the deflection rises with α increasing by a factor of 2.8; 7.2; 13.3 for $\alpha = 0.1$; 0.3; 0.5, respectively, in comparison with the deflection for $\alpha = 0$.

In a shell with corrugation frequency $k = 8$ the deflection of the crest portion of corrugation ($\varphi = 0$) is insignificant when $\alpha = 0.1$. When $\alpha = 0.3$, the trough portion deflects a little against the applied load and at $\alpha = 0.5$ coincides with it. The convex portion of corrugation ($\varphi = \pi/4$) deflects with the applied load for all values of the

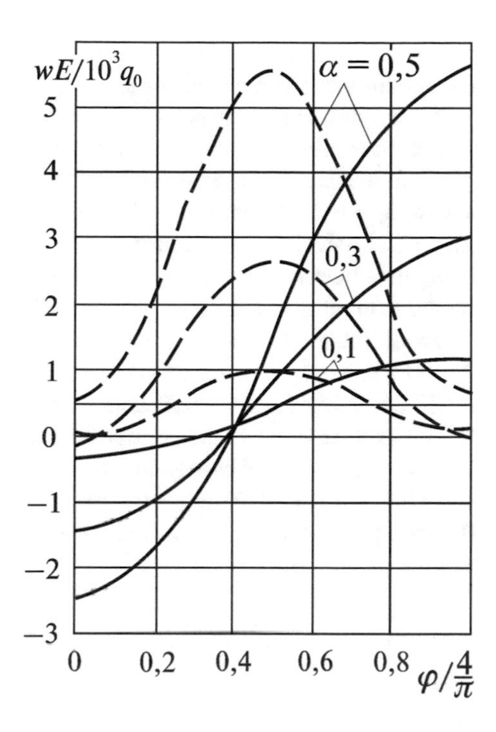

Fig. 3.3 Distribution of the deflection w

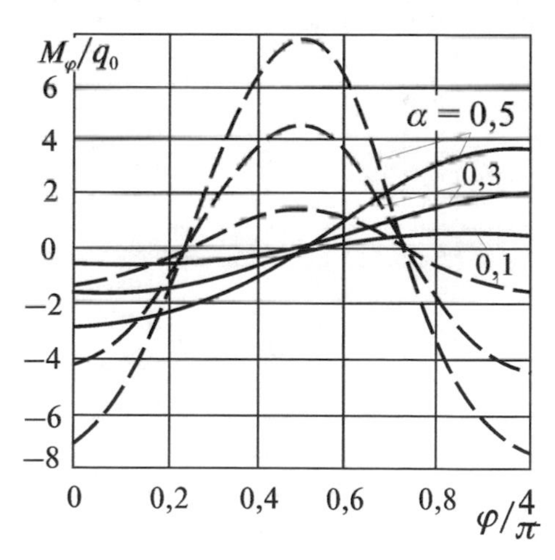

Fig. 3.4 Distribution of the bending moment M_φ

parameter α and this deflection increases with α by factors of 2.3; 6.4 and 12.8 at $\alpha = 0.1$, 0.3, and 0.5, respectively, compared with the deflection at $\alpha = 0$.

From Fig. 3.4 it can be seen how the bending moment varies at $k = 4$ and $k = 8$ on the intervals [0, π/4] and [0, π/8]. As the corrugation amplitude increases, the maximum moment increases by factors of 3.3 and 6.2 for $k = 4$ ($\varphi = \pi/4$) and by factors of 3.1 and 5.3 for $k = 8$ ($\varphi = \pi/8$) and $\alpha = 0.3$ and 0.5, respectively, when compared with the moment for $\alpha = 0.1$.

We now consider a shell, whose thickness varies according to Eq. (3.2.20) and whose ends are rigidly fixed. The shell is subjected to a load q_0. For solving the problem we have used the following input data: $r_0 = 15$, $L = 30$, $h_0 = 0.5$, $\nu = 0.3$, $k = 4$, 8, $\alpha = 0.1$. The corresponding values of the parameter β are presented in Table 2.11, which shows the distribution of the deflection w and bending moment M_φ along the directrix at $s = L/2 = 15$ for five points (numbered by n and evenly spaced) on the interval $\varphi = \pi/8$ for $k = 4$. The value $k = 0$ corresponds to a circular shell.

From Table 3.8 it is seen that the maximum deflection in the circular shell at $\beta > 0$ increases by a factor of 1.3 and 2 for $\beta = 0.25$ and 0.5 and at $\beta < 0$ decreases by 20 and 33 % for $\beta = -0.25$ and -0.5, compared to the maximum deflection for $\beta = 0$. As β increases in the corrugated shell with $k = 4$, the maximum deflection of the through portion of corrugation ($\varphi = \pi/4$) increases in absolute magnitude by 17 and 49 % for $\beta = -0.25$ and -0.5 and decreases by 10 and 15 % for $\beta = 0.25$ and 0.5, compared with the deflection for $\beta = 0$. The corrugation crest ($\varphi = 0$) moves against the applied load.

Table 3.8 Maximum deflection in the circular shell at $\beta > 0$

$Ew/(10^3 q_0)$						
k	n	$\beta = 0$	$\beta = 0.25$	$\beta = -0.25$	$\beta = 0.5$	$\beta = -0.5$
0	1–5	0.4138	0.5519	0.311	0.267	0.767
4	1	−0.2480	−0.2513	−0.415	−0.605	−0.287
	2	−0.0442	0.177	−0.801	0.301	−0.979
	3	0.4505	0.285	0.882	0.393	0.362
	4	0.9490	0.095	1.189	0.080	1.293
	5	1.1565	1.413	1.522	0.868	1.223
8	1	0.0573	0.204	0.091	−0.206	0.896
	2	0.1909	0.089	0.064	0.736	0.601
	3	0.5121	0.713	0.817	0.776	0.870
	4	0.8308	0.458	0.394	0.063	0.8802
	5	0.9620	0.417	1.234	0.765	1.1625
M_ϕ/q_0						
0	1–5	0	0.0003	−0.0029	0	−0.0082
4	1	−0.5816	−0.3485	−0.8638	−0.1737	−1.1788
	2	−0.4186	−0.1843	−0.6987	−0.0105	−1.0048
	3	−0.0086	0.2059	−0.2622	0.3529	−0.5277
	4	0.4246	0.5864	0.2248	0.6671	0.0200
	5	0.6109	0.7408	0.4404	0.7810	0.2649
8	1	−1.4699	−0.9419	−2.0001	−0.5011	−2.4394
	2	−1.0373	−0.5050	−1.5701	−0.0570	−2.0118
	3	0.0061	0.5462	−0.5308	1.0093	−0.9768
	4	1.0472	1.5909	0.5093	2.0632	0.0611
	5	1.4778	2.0216	0.9402	2.4947	0.4915

This deflection increases in absolute value by 1.3 and 5 % for $\beta = 0.25$ and 0.5 and decreases by 3 and 8 % for $\beta = -0.25$ and -0.5, compared to the deflection for $\beta = 0$. In the corrugated shell with k = 8, the corrugation crest moves with the load when $\beta = 0, \pm 0.25, -0.5$ and against the load when $\beta = 0.5$.

The deflection of the crest portion increases in absolute value by factors of 1.9 and 3.3 for $\beta = -0.25$ and -0.5, respectively, compared with the deflection for $\beta = 0$.

The maximum deflection of the trough portion increases by 6 and 21 % for $\beta = -0.25$ and -0.5 and by 1.5 % for $\beta = 0.5$ and decreases by 2 % for $\beta = 0.25$, compared to the deflection for $\beta = 0$. For k = 4 the maximum deflection exceeds the one for the circular shell by a factor of 2.8 and for k = 8 by a factor of 1.2.

From the table it follows that, when k = 4, the moment M for $\beta = 0$ is negative and increases in absolute value by factors of 1.5 and 2 for $\beta = -0.25$ and -0.5 and decreases by factors of 1.7 and 3.3 for $\beta = 0.25$ and 0.5, compared with the moment for $\beta = 0$. When $\beta \geq 0$, the moment is maximum for $\varphi = \pi/4$ and increases by 21 and 28 % for $\beta = 0.25$ and 0.5. When $\beta < 0$, the moment is maximum for $\varphi = 0$.

When $k = 8$, the moment is maximum for $\varphi = \pi/8$ and increases by 37 and 69 % for $\beta = 0.25$ and 0.5, and, when $\phi = 0$, the moment increases in absolute value by 36 and 69 % for $\beta = 0.25$ and -0.5, compared with the moment for $\beta = 0$.

In order to study the effect of the corrugation amplitude on the stress state of shells with the thickness varying along the directrix, we solved the problem with the following data: $r_0 = 15$; $L = 30$; $h_0 = 0.5$; $k = 8$; $\alpha = 0$; 0.1; 0.3; 0.5; $\beta = 0$; $\pm$ 0.25; $\pm$ 0.5; $\pm$ 0.75. The values of the deflection w on the interval $0 \leq \varphi \leq \pi/8$ in the section s = L/2 are presented in Table 3.9.

Table 3.9 Values of the deflection w

$Ew/(10^3 q_0)$

α	$\varphi/\frac{\pi}{8}$	$\beta = 0$	$\beta = 0.25$	$\beta = -0.25$	$\beta = 0.5$	$\beta = -0.5$	$\beta = 0.75$	$\beta = -0.75$
0.1	0	0.0573	0.0204	0.1092	−0.0206	0.1896	−0.1574	0.3483
	0.25	0.1909	0.2089	0.2064	0.2736	0.2601	0.4679	0.3988
	0.5	0.5121	0.5713	0.4817	0.6776	0.4869	0.9323	0.5787
	0.75	0.8308	0.8458	0.8394	0.9063	0.8802	1.1182	1.0154
	1	0.9620	0.9417	1.0234	0.9765	1.1625	1.1694	1.5919
0.3	0	−0.0896	−0.2286	0.0435	−0.4373	0.2019	−1.0480	0.4713
	0.25	0.3246	0.3532	0.3462	0.4656	0.4226	0.8513	0.6311
	0.5	1.3189	1.4686	1.2028	1.6993	1.1321	2.2480	1.2001
	0.75	2.2922	2.3006	2.3005	2.3846	2.3445	2.7932	2.5620
	1	2.6853	2.5855	2.8545	2.5906	3.1979	2.9398	4.3211
0.5	0	0.5505	0.2841	0.8001	−0.0946	1.0999	−1.0842	1.6697
	0.25	1.2688	1.2836	1.3319	1.4457	1.4937	2.1210	1.9604
	0.5	3.0065	3.2079	2.8521	3.5508	2.7747	4.1648	3.0089
	0.75	4.6869	4.6198	4.7838	4.6941	4.9500	5.3501	5.4999
	1	5.3478	5.0891	5.7354	5.0263	6.4467	5.5791	8.6356

From the table it is seen that the corrugation crest ($\varphi = 0$) with $\alpha = 0.1$ moves with the applied load when $\beta = 0, \pm 0.25, -0.5, -0.75$ and against the load, when $\beta = 0.5$ and 0.75.

The maximum deflection of the trough portion of corrugation ($\varphi = \pi/8$) increases by 6, 21 and 65 % for $\beta = -0.25$; -0.5, and -0.75; by 1.5 and 22 % for $\beta = 0.5$ and 0.75, and decreases by 2 % for $\beta = 0.25$, compared with the deflection for $\beta = 0$. The corrugation crest in the shell with $\alpha = 0.3$ moves with the load when $\beta < 0$, and against the load, when $\beta \geq 0$.

The deflection increases in absolute value by factors of 2.6; 4.9, and 11.7 for $\beta = 0.25$; 0.5, and 0.75 and by factors of 2.3 and 5.3 for $\beta = -0.5$ and -0.75, compared with the deflection for $\beta = 0$.

The maximum increases by 6, 15, and 61 % for $\beta = -0.25$; -0.5, and -0.75 and by 9 % for $\beta = 0.75$, compared with the deflection for $\beta = 0$. The corrugation crest in the shell with $\alpha = 0.5$ moves with the applied load for $\beta = 0$; $\pm$ 0.25; -0.5; -0.75 and against the load for $\beta = 0.5$ and 0.75. The maximum deflection increases by 7.2; 21, and 61 % for $\beta = -0.25$; -0.5, and -0.75, respectively, compared with the deflection for $\beta = 0$.

Table 3.10 presents values of the bending moment M_φ on the interval $0 \leq \varphi \leq \pi/8$ in the section $s = L/2$. From the table it is seen that, when $\alpha = 0.1$, the moment is maximum for $\varphi = \pi/8$ with $\beta > 0$ and for $\beta < 0$ with $\varphi = 0$.

The maximum moment increases with β by 37, 69, and 94 %, when $\beta > 0$, and by 35, 65, and 84 %, when $\beta < 0$, for $\beta = \pm 0.25$; ± 0.5; ± 0.75, compared with the moment for $\beta = 0$.

Table 3.10 Values of the bending moment

M_φ/q_0								
α	$\varphi/\frac{\pi}{8}$	$\beta = 0$	$\beta = 0.25$	$\beta = -0.25$	$\beta = 0.5$	$\beta = -0.5$	β 0.75	$\beta = -0.75$
0.1	0	−1.4699	−0.9419	−2.0001	−0.5011	−2.4394	−0.1892	−2.7247
	0.25	−1.0373	−0.5050	−1.5701	−0.0570	−2.0118	0.2683	−2.3006
	0.5	0.0061	0.5462	−0.5308	1.0094	−0.9768	1.3647	−1.2760
	0.75	1.0472	1.5910	0.5093	2.0632	0.06109	2.4361	−0.2486
	1	1.4778	2.0216	0.9402	2.4947	0.4915	2.8682	0.1779
0.3	0	−4.4304	−2.8274	−6.0523	−1.4977	−7.4173	−0.5603	−8.3703
	0.25	−3.1184	−1.5035	−4.7465	−1.5626	−6.1148	0.8045	−7.0732
	0.5	0.0505	1.6835	−1.5857	3.0621	−2.9580	4.0670	−3.9237
	0.75	3.2128	4.8476	1.5765	6.2321	0.2044	7.2425	−0.7590
	1	4.5183	6.1485	2.8836	7.5256	1.5123	8.5254	0.5528
0.5	0	−7.4318	−4.7053	−10.268	−2.4799	−12.743	−0.9198	−14.548
	0.25	−5.2163	−2.4736	−8.0628	−0.2305	−10.545	1.3351	−12.357
	0.5	0.1665	2.9087	−2.6657	5.1602	−5.1263	6.7077	6.9224
	0.75	5.5639	8.2436	2.7983	10.435	0.4019	11.886	1.3357
	1	7.7933	10.429	5.0683	12.571	2.7076	13.950	1.0335

When $\alpha = 0.3$ the maximum moment increases by 36, 67, and 88 % for $\beta = 0.25$; 0.5, and 0.75 and by 34, 64 and 84 % for $\beta = -0.25$; −0.5, and −0.75, compared with the moment for $\beta = 0$.

When $\alpha = 0.5$, the maximum moment increases by 34, 61 and 79 % for $\beta = 0.25$; 0.5, and 0.75 and by 31, 63, and 87 % for $\beta = -0.25$; −0.5 and −0.75, compared with the moment for $\beta = 0$.

Thus, from the results presented in the figures and tables, it can be seen how the geometric characteristics (frequency and amplitude) of a corrugated shell affect the redistribution of factors of its stress-strain state.

3.3 Stress-Strain State of Shells of Revolution

3.3.1 Governing Equations

In general, we will consider laminated shells of revolution composed of the arbitrary number of isotropic and anisotropic layers with the thickness varying along a generatrix, so that the bounding surfaces and contact surfaces of adjacent layers be aligned surfaces of revolution (Flügge [3], Grigorenko [6], Johns [11], Li et al. [13], Sun et al. [19], Zang [21]).

When choosing in a shell a coordinate surface, we characterize it by a curvilinear orthogonal system of coordinates, $\alpha = s$, $\beta = \theta$, where s is the length of a meridian arc and θ is the central angle in a parallel circle (Fig. 3.5). The lines s = const and θ = const are the lines of principal curvatures. By orienting the coordinate γ along the normal to this surface, we describe the whole shell by an orthogonal spatial coordinate system s, θ, γ.

In the case when the shell is a single-layer and composed of layers of variable thickness, we assume that the coordinate surface coincides with the median one. If the layers are of constant thickness, then the coordinate surface is equidistant with respect to interfaces of adjacent layers and bounding surfaces.

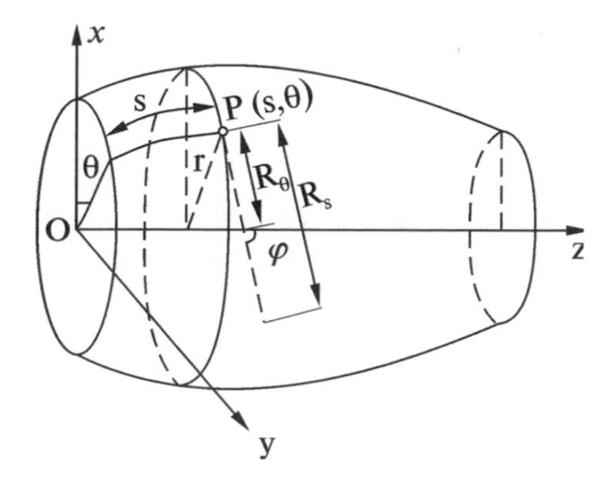

Fig. 3.5 Shell of revolution

The contact and bounding surfaces in the case of layers with variable thickness will be analyzed in other coordinate systems, which may be oblique.

The equations of the coordinate surface in the orthogonal coordinate system x, y, z, whose Oz-axis coincides with the axis of revolution (Fig. 3.5), are:

$$\begin{gathered} x = r(s)\cos\theta, \quad y = r(s)\sin\theta, \quad z = z(s) \\ s_0 \leq s \leq s_N, \quad -\pi \leq \theta_1 \leq \theta \leq \theta_2 \leq \pi, \end{gathered} \tag{3.3.1}$$

where $r(s)$ is the radius of the parallel circle; $z(s)$ is the distance from the initial plane $z = z_0$ along the axis of revolution.

Suppose that functions $r(s)$ and $z(s)$ are continuous and have a sufficient number of piecewise continuous derivatives on the interval $s_0 \leq s \leq s_N$. By taking (3.3.1) into account and by using (1.2.13)–(1.2.18), we obtain expressions for the first and second quadratic forms (Korn and Korn [12]):

$$\varphi_1 = \mathrm{d}s^2 + r^2\mathrm{d}\theta^2, \tag{3.3.2}$$

and:

$$\varphi_2 = -\left(\frac{1}{R_s}\mathrm{d}s^2 + \frac{r^2}{R_\theta}\mathrm{d}\theta^2\right), \tag{3.3.3}$$

where R_s, R_θ are the radiuses of principal curvatures in the meridional and circumferential directions, respectively.

The Gauss-Codazzi relations (1.2.19) reduce to a single equation:

$$\frac{\mathrm{d}r}{\mathrm{d}s} = \cos\varphi, \tag{3.3.4}$$

where φ is the angle between the normal to the coordinate surface and the axis of revolution.

In view of (3.3.4), the curvature radii R_s and R_θ can be presented as:

$$R_s = -\frac{\sqrt{1-(r')^2}}{r''}, \quad R_\theta = \frac{r}{\sqrt{1-(r')^2}}. \tag{3.3.5}$$

From relations of Chap. 1, acknowledging for (3.3.1)–(3.3.5) and omitting nonlinear terms, we obtain the following equations for shells of revolution;

for the rotation angles of the normal:

$$N, \vartheta_\theta = -\frac{1}{r}\frac{\partial w}{\partial \theta} + \frac{v}{R_\theta}; \tag{3.3.6}$$

for the strain components of the coordinate (median) surface:

$$
\begin{aligned}
&\varepsilon_s = \frac{\partial u}{\partial s} + \frac{w}{R_s}, \quad \varepsilon_\theta = \frac{1}{r}\frac{\partial v}{\partial \theta} + \frac{\cos\varphi}{r}u + \frac{\sin\varphi}{r}w, \\
&\varepsilon_{s\theta} = \frac{1}{r}\frac{\partial u}{\partial \theta} + \frac{\partial v}{\partial s} - \frac{\cos\varphi}{r}v; \quad \kappa_s = \frac{\partial \vartheta_s}{\partial s},
\end{aligned} \tag{3.3.7}
$$

and:

$$
\begin{aligned}
&\kappa_\theta = \frac{1}{r^2}\left(\sin\varphi\frac{\partial v}{\partial \theta} - \frac{\partial^2 w}{\partial \theta^2}\right) + \frac{\cos\varphi}{r}\vartheta_s, \\
&\kappa_{s\theta} = \frac{\sin\varphi}{r}\left(\frac{\partial v}{\partial s} - \frac{\cos\varphi}{r}v\right) + \frac{\cos\varphi}{r^2}\frac{\partial w}{\partial \theta} + \frac{1}{r}\frac{\partial \vartheta_s}{\partial \theta};
\end{aligned} \tag{3.3.8}
$$

equilibrium equations:

$$
\begin{aligned}
&\frac{\partial}{\partial s}(rN_s) + \frac{\partial S}{\partial \theta} - \cos\varphi N_\theta + \frac{1}{R_s}\frac{\partial H}{\partial \theta} + \frac{r}{R_s}Q_s + rq_s = 0, \\
&\frac{\partial}{\partial s}(rS) + \frac{\partial N_\theta}{\partial \theta} - \cos\varphi S + \frac{\partial}{\partial s}(\sin\varphi H) + \frac{\cos\varphi}{R_s}H + \sin\varphi Q_\theta + rq_\theta = 0, \\
&\frac{\partial}{\partial s}(rQ_s) + \frac{\partial Q_\theta}{\partial \theta} - \frac{r}{R_s}N_s - \sin\varphi \quad N_\theta + rq_\gamma = 0, \\
&\frac{\partial}{\partial s}(rM_s) + \frac{\partial H}{\partial \theta} - \cos\varphi M_\theta - rQ_s = 0, \\
&\frac{\partial}{\partial s}(rH) + \frac{\partial M_\theta}{\partial \theta} + \cos\varphi H - rQ_\theta = 0.
\end{aligned} \tag{3.3.9}
$$

The elasticity relations have the form (1.4.11). Boundary conditions can be specified in terms of forces and moments as follows:

$$
\begin{aligned}
&N_s, M_s, \hat{S} = S + \frac{2\sin\varphi}{r}H, \quad \hat{Q}_s = Q_s + \frac{1}{r}\frac{\partial H}{\partial \theta}, \quad \text{for} \quad s = \text{const.}, \\
&N_\theta, M_\theta, \hat{S} = S + \frac{2}{R_s}H, \quad \hat{Q}_\theta = Q_\theta + \frac{\partial H}{\partial s} \quad \text{for} \quad \theta = \text{const.},
\end{aligned} \tag{3.3.10}
$$

or in terms of displacements as

$$
\begin{aligned}
&u, v, w, \vartheta_\alpha \quad \text{for} \quad s = \text{const}, \\
&u, v, w, \vartheta_\beta \quad \text{for} \quad \theta = \text{const}.
\end{aligned} \tag{3.3.11}
$$

Let us consider shells of revolution that are closed in a circumferential direction. Then, due to the periodicity of load components, contour values, and stress-strain parameters, these factors can be expanded into Fourier series in the coordinate θ:

$$
\begin{aligned}
q_s(s,\theta) &= \sum_{k=0}^{\infty}\left[q_{s,k}(s)\cos k\theta + q'_{s,k}(s)\sin k\theta\right],\\
q_\theta(s,\theta) &= \sum_{k=0}^{\infty}\left[q_{\theta,k}(s)\sin k\theta + q'_{\theta,k}(s)\cos k\theta\right],\\
q_\gamma(s,\theta) &= \sum_{k=0}^{\infty}\left[q_{\gamma,k}(s)\cos k\theta + q'_{\gamma,k}(s)\sin k\theta\right],
\end{aligned}
\tag{3.3.12}
$$

$$
\begin{aligned}
N_s(s,\theta) &= \sum_{k=0}^{\infty}\left[N_{s,k}(s)\cos k\theta + N'_{s,k}(s)\sin k\theta\right],\\
N_\theta(s,\theta) &= \sum_{k=0}^{\infty}\left[N_{\theta,k}(s)\cos k\theta + N'_{\theta,k}(s)\sin k\theta\right],\\
S(s,\theta) &= \sum_{k=0}^{\infty}\left[S_k(s)\sin k\theta + S'_k(s)\cos k\theta\right],\\
M_s(s,\theta) &= \sum_{k=0}^{\infty}\left[M_{s,k}(s)\cos k\theta + M'_{s,k}(s)\sin k\theta\right],\\
M_\theta(s,\theta) &= \sum_{k=0}^{\infty}\left[M_{\theta,k}(s)\cos k\theta + M'_{\theta,k}(s)\sin k\theta\right],\\
H(s,\theta) &= \sum_{k=0}^{\infty}\left[H_k(s)\sin k\theta + H'_k(s)\cos k\theta\right],\\
Q_s(s,\theta) &= \sum_{k=0}^{\infty}\left[Q_{s,k}(s)\cos k\theta + Q'_{s,k}(s)\sin k\theta\right],\\
Q_\theta(s,\theta) &= \sum_{k=0}^{\infty}\left[Q_{\theta,k}(s)\sin k\theta + Q'_{\theta,k}(s)\cos k\theta\right],\\
\varepsilon_s(s,\theta) &= \sum_{k=0}^{\infty}\left[\varepsilon_{s,k}(s)\cos k\theta + \varepsilon'_{s,k}(s)\sin k\theta\right]\\
\varepsilon_\theta(s,\theta) &= \sum_{k=0}^{\infty}\left[\varepsilon_{\theta,k}(s)\cos k\theta + \varepsilon_{\theta,k}(s)\sin k\theta\right],\\
\varepsilon_{s\theta}(s,\theta) &= \sum_{k=0}^{\infty}\left[\varepsilon_{s\theta,k}(s)\sin k\theta + \varepsilon_{s\theta,k}(s)\cos k\theta\right],
\end{aligned}
\tag{3.3.13}
$$

and:

$$\begin{aligned}
\kappa_s(s,\theta) &= \sum_{k=0}^{\infty}\left[\kappa_{s,k}(s)\cos k\theta + \kappa'_{s,k}(s)\sin k\theta\right],\\
\kappa_\theta(s,\theta) &= \sum_{k=0}^{\infty}\left[\kappa_{\theta,k}(s)\cos k\theta + \kappa'_{\theta,k}(s)\sin k\theta\right],\\
\kappa_{s\theta}(s,\theta) &= \sum_{k=0}^{\infty}\left[\kappa_{s\theta,k}(s)\sin k\theta + \kappa'_{s\theta,k}(s)\cos k\theta\right],\\
u(s,\theta) &= \sum_{k=0}^{\infty}\left[u_k(s)\cos k\theta + u'_k(s)\sin k\theta\right],\\
v(s,\theta) &= \sum_{k=0}^{\infty}\left[v_k(s)\sin k\theta + v_k(s)\cos k\theta\right],\\
w(s,\theta) &= \sum_{k=0}^{\infty}\left[w_k(s)\cos k\theta + w_k(s)\sin k\theta\right].
\end{aligned} \tag{3.3.14}$$

After substituting (3.3.12)–(3.3.14) into the relations (3.3.6)–(3.3.10) and (1.4. 11), and after collecting the corresponding terms in $\cos k\theta$ and $\sin k\theta$, we obtain the following dependencies for the amplitude values (without primes), for the strains expressed in terms of displacements:

$$\begin{aligned}
\varepsilon_s &= \frac{du}{ds} + \frac{w}{R_s}, \quad \varepsilon_\theta = \frac{1}{r}(kv + \cos\varphi u + \sin\varphi w),\\
\varepsilon_{s\theta} &= -\frac{k}{r}u + \frac{dv}{ds} - \frac{\cos\varphi}{r}v,
\end{aligned} \tag{3.3.15}$$

and:

$$\begin{aligned}
\kappa_s &= \frac{d\vartheta_s}{ds}, \quad \kappa_\theta = \frac{k}{r^2}(kw + \sin\varphi v) + \frac{\cos\varphi}{r}\vartheta_s,\\
\kappa_{s\theta} &= -\frac{k}{r^2}(r\vartheta_s + \cos\varphi w) + \frac{\sin\varphi}{r}\left(\frac{dv}{ds} - \frac{\cos\varphi}{r}v\right),
\end{aligned} \tag{3.3.16}$$

where:

$$\vartheta_s = -\frac{dw}{ds} + \frac{u}{R_s}; \tag{3.3.17}$$

equilibrium equations:

$$
\begin{aligned}
&\frac{\mathrm{d}rN_s}{\mathrm{d}s} + kS - \cos\varphi N_\theta + \frac{k}{R_s}H + r\frac{Q_s}{R_s} + rq_s = 0,\\
&\frac{\mathrm{d}rS}{\mathrm{d}s} - kN_\theta + \cos\varphi S + \frac{\mathrm{d}\sin\varphi H}{\mathrm{d}s} + \frac{\cos\varphi}{R_s}H + \sin\varphi Q_\theta + rq_\vartheta = 0,\\
&\frac{\mathrm{d}rQ_s}{\mathrm{d}s} + kQ_\theta - \frac{r}{R_s}N_s - \sin\varphi N_\theta + rq_\gamma = 0,\\
&\frac{\mathrm{d}rM_s}{\mathrm{d}s} + kH - \cos\varphi M_\theta - rQ_s = 0,\\
&\frac{\mathrm{d}rH}{\mathrm{d}s} - kM_\theta + \cos\varphi H - rQ_\theta = 0.
\end{aligned}
\tag{3.3.18}
$$

The elasticity relations retain the form (1.22). The same dependencies can be obtained for the amplitude values with primes by replacing k in (3.3.15)–(3.3.18) by $-k$.

When using the general approach, it is necessary to take those requirements into account that follow from the method by which the problem under consideration is solved. In this case it is expedient to choose base functions such that conditions for combination of different shells of revolution can be realized by the simplest method. The intention is to derive the resolving equations and later on construct the algorithm for solving the problem for shells of revolution of arbitrary form as well as for the system composed of shells with different shape.

In order to satisfy the requirements above, we choose the following functions as base ones:

$$
N_r, N_z, \hat{S}, M_s, u_r, u_z, v, \vartheta_s, \tag{3.3.19}
$$

where:

$$
N_r = \cos\varphi N_s + \sin\varphi \hat{Q}_s, \quad N_z = \sin\varphi N_s + \cos\varphi \hat{Q}_s, \tag{3.3.20}
$$

and:

$$
u_r = \cos\varphi u + \sin\varphi w, \quad u_z = \sin\varphi u - \cos\varphi w. \tag{3.3.21}
$$

N_r and N_z are the radial and axial forces, and u_r and u_z the corresponding displacements, respectively.

Let us, in general, consider the stress-strain state of a closed multilayer shell of revolution compounded of an arbitrary number of anisotropic layers of variable thickness, assuming at the same time that the shell at each point has one plane of elastic symmetry, passing in parallel to its coordinate surface.

After some transformations of the amplitude values (k = 1, 2, …), we arrive at the resolving system of ordinary differential equations in the normal Cauchy form of the sixteenth order:

$$\frac{d\bar{N}}{ds} = A(s)\bar{N} + \bar{f}(s), \tag{3.3.22}$$

where:

$$\bar{N} = \{N_r, N_z, \hat{S}, M_s, u_r, u_z, v, \vartheta_s, N'_r, N'_z, \ldots, \vartheta'_s\}, \tag{3.3.23}$$

and:

$$A(s) = \|a_{ij}(s)\| \quad (i, j = 1, 2, \ldots, 16). \tag{3.3.24}$$

Moreover:

$$\bar{f} = \{f_1, f_2, \ldots, f_{16}\}. \tag{3.3.25}$$

The matrix $A(s)$ can be written as:

$$A(s) = \begin{Vmatrix} A_1 & A_2 \\ A_3 & A_4 \end{Vmatrix}, \tag{3.3.26}$$

where A_t (t = 1, 2, 3, 4) are square matrices of the eighth order. The matrix A_4 can be derived from A_1 by changing the signs in the columns and lines at $\hat{S}$ and v. The matrix A_2 may be derived from A_3 in the same way. The elements of the matrices A_1 and A_2 and the components of the vector $\bar{f}(s)$ are explicitly given by:

$$a_{1,1} = -\frac{\cos\varphi}{r} + \frac{d_{41}\cos\varphi}{r} + k^2\frac{d_{51}\sin\varphi\cos\varphi}{r^2},$$

$$a_{1,2} = \frac{d_{41}\sin\varphi}{r} + k^2\frac{d_{51}\sin^2\varphi}{r^2}, \quad a_{1,3} = -k\frac{\cos\varphi}{r}, \quad a_{1,4} = \frac{d_{43}}{r} + k^2\frac{d_{53}\sin\varphi}{r^2},$$

$$a_{1,5} = \frac{d_{44}}{r^2} + k^2\frac{(d_{45}+d_{54})\sin\varphi}{r^3} + k^4\frac{d_{55}\sin^2\varphi}{r^4},$$

$$a_{1,6} = -k^2\frac{d_{45}\cos\varphi}{r^3} - k^4\frac{d_{55}\sin\varphi\cos\varphi}{r^4},$$

$$a_{1,7} = k\frac{d_{44}r + d_{45}\sin\varphi}{r^3} + k^3\frac{(d_{54}r + d_{55}\sin\varphi)\sin\varphi}{r^4},$$

$$a_{1,8} = \frac{d_{45}\cos\varphi}{r^2} + k^2\frac{d_{55}\sin\varphi\cos\varphi}{r^3},$$

$$a_{1,11} = \frac{d_{42}}{r} + k^2\frac{d_{52}\sin\varphi}{r^2}, \quad a_{1,14} = -k\frac{d_{46}}{r^3} - k^3\frac{d_{56}\sin\varphi}{r^4},$$

$$a_{1,16} = k\frac{d_{46}}{r^2} + k^3\frac{d_{56}\sin\varphi}{r^3}, \quad a_{1,9} = a_{1,10} = a_{1,12} = a_{1,13} = a_{1,15} = 0,$$

$$a_{2,1}=-k^2\frac{d_{51}\cos^2\varphi}{r^2},\quad a_{2,2}=-\frac{\cos\varphi}{r}-k^2\frac{d_{51}\sin\varphi\cos\varphi}{r^2},$$
$$a_{2,3}=k\frac{2d_{62}-r\sin\varphi}{r^2},\quad a_{2,4}=-k^2\frac{d_{53}\cos\varphi}{r^2},$$
$$a_{2,5}=-k^2\frac{d_{54}\cos\varphi}{r^3}-k^4\frac{d_{55}\sin\varphi\cos\varphi}{r^4},\quad a_{2,6}=k^2\frac{2d_{66}}{r^4}+k^4\frac{d_{55}\cos^2\varphi}{r^4},$$
$$a_{2,7}=-k^3\frac{(d_{54}r+d_{55}\sin\varphi)\cos\varphi}{r^4},\quad a_{2,8}=-k^2\frac{2d_{66}+d_{55}\cos^2\varphi}{r^3},$$
$$a_{2,9}=k\frac{2d_{61}\cos\varphi}{r^2},\quad a_{2,10}=k\frac{2d_{61}\sin\varphi}{r^2},\quad a_{2,11}=-k^2\frac{d_{52}\cos\varphi}{r^2},$$
$$a_{2,12}=k\frac{2d_{63}}{r^2},\quad a_{2,13}=k\frac{2d_{64}}{r^3}+k^3\frac{2d_{65}\sin\varphi}{r^4},$$
$$a_{2,14}=k^3\frac{(d_{56}-2d_{65})\cos\varphi}{r^4},$$
$$a_{2,15}=-k^2\frac{2(d_{64}r+d_{65}\sin\varphi)}{r^4},\quad a_{2,16}=k\frac{2d_{65}\cos\varphi}{r^3}-k^3\frac{d_{56}\cos\varphi}{r^3},$$

$$a_{3,1}=k\frac{(d_{41}r+d_{51}\sin\varphi)\cos\varphi}{r^2},\quad a_{3,2}=k\frac{(d_{41}r+d_{51}\sin\varphi)\sin\varphi}{r^2},$$
$$a_{3,3}=-\frac{2\cos\varphi}{r},\quad a_{3,4}=k\frac{d_{43}r+d_{53}\sin\varphi}{r^2},$$
$$a_{3,5}=k\frac{d_{44}r+d_{54}\sin\varphi}{r^3}+k^3\frac{(d_{45}r+d_{55}\sin\varphi)\sin\varphi}{r^4},$$
$$a_{3,6}=-k^3\frac{(d_{45}r+d_{55}\sin\varphi)\cos\varphi}{r^4},$$
$$a_{3,7}=k^2\frac{(d_{44}r+d_{45}\sin\varphi)}{r^3}+\frac{(d_{54}r+d_{55}\sin\varphi)\sin\varphi}{r^4},$$
$$a_{3,8}=k\frac{(d_{45}r+d_{55}\sin\varphi)\cos\varphi}{r^3},\quad a_{3,11}=k\frac{d_{42}r+d_{52}\sin\varphi}{r^2},$$
$$a_{3,14}=-k^2\frac{d_{46}r+d_{56}\sin\varphi}{r^4},\quad a_{3,16}=k^2\frac{d_{46}r+d_{56}\sin\varphi}{r^3},$$
$$a_{3,9}=a_{3,10}=a_{3,12}=a_{3,13}=a_{3,15}=0,\quad a_{4,1}=\sin\varphi+\frac{d_{51}\cos^2\varphi}{r},$$
$$a_{4,2}=-\cos\varphi+\frac{d_{51}\sin\varphi\cos\varphi}{r},\quad a_{4,3}=-k\frac{2d_{62}}{r},\quad a_{4,4}=\frac{-\cos\varphi+d_{53}\cos\varphi}{r},$$

$$a_{4,5}=\frac{d_{54}\cos\varphi}{r^2}+k^2\frac{d_{55}\sin\varphi\cos\varphi}{r^3},\quad a_{4,6}=-k^2\frac{d_{55}\cos^2\varphi+2d_{66}}{r^3},\tag{3.3.27}$$

$$a_{4,7} = k\frac{(d_{54}r + d_{55}\sin\varphi)\cos\varphi}{r^3}, \quad a_{4,8} = \frac{d_{55}\cos^2\varphi + k^2 2d_{66}}{r^2},$$

$$a_{4,9} = -k\frac{2d_{61}\cos\varphi}{r}, \quad a_{4,10} = -k\frac{2d_{61}\sin\varphi}{r}, \quad a_{4,11} = \frac{d_{52}\cos\varphi}{r},$$

$$a_{4,12} = -k\frac{2d_{63}}{r}, \quad a_{4,13} = -k\frac{2d_{64}}{r^2} - k^3\frac{2d_{65}\sin\varphi}{r^3},$$

$$a_{4,14} = -k\frac{d_{56}\cos\varphi}{r^3} + k^3\frac{2d_{65}\cos\varphi}{r^3},$$

$$a_{4,15} = k^2\frac{2(d_{64}r + d_{66}\sin\varphi)}{r^3}, \quad a_{4,16} = k\frac{d_{56}\cos\varphi - 2d_{65}\cos\varphi}{r^2},$$

$$a_{5,1} = d_{11}\cos^2\varphi, \quad a_{5,2} = d_{11}\sin\varphi\cos\varphi, \quad a_{5,4} = d_{13}\cos\varphi,$$

$$a_{5,5} = \frac{d_{14}\cos\varphi}{r} + k^2\frac{d_{15}\sin\varphi\cos\varphi}{r^2}, \quad a_{5,6} = -k^2\frac{d_{15}\cos^2\varphi}{r^2},$$

$$a_{5,7} = k\frac{(d_{14}r + d_{15}\sin\varphi)\cos\varphi}{r^2}, \quad a_{5,8} = -\sin\varphi + \frac{d_{15}\cos^2\varphi}{r},$$

$$a_{5,11} = d_{12}\cos\varphi, \quad a_{5,14} = -k\frac{d_{16}\cos\varphi}{r^2}, \quad a_{5,16} = k\frac{d_{16}\cos\varphi}{r},$$

$$a_{5,3} = a_{5,9} = a_{5,10} = a_{5,12} = a_{5,13} = a_{5,15} = 0,$$

$$a_{6,1} = d_{11}\sin\varphi\cos\varphi, \quad a_{6,2} = d_{11}\sin^2\varphi, \quad a_{6,4} = d_{12}\sin\varphi,$$

$$a_{6,5} = \frac{d_{13}\sin\varphi}{r} + k^2\frac{d_{14}\sin^2\varphi}{r^2}, \quad a_{6,6} = -k^2\frac{d_{15}\sin\varphi\cos\varphi}{r^2},$$

$$a_{6,7} = k\frac{(d_{14}r + d_{15}\sin\varphi)\sin\varphi}{r^2}, \quad a_{6,8} = \cos\varphi + \frac{d_{15}\sin\varphi\cos\varphi}{r},$$

$$a_{6,11} = d_{12}\sin\varphi, \quad a_{6,14} = -k\frac{d_{16}\sin\varphi}{r^2}, \quad a_{6,16} = k\frac{d_{16}\sin\varphi}{r},$$

$$a_{6,3} = a_{6,9} = a_{6,10} = a_{6,12} = a_{6,13} = a_{6,15} = 0,$$

$$a_{7,3} = d_{22}, \quad a_{7,5} = k\frac{\cos\varphi}{r}, \quad a_{7,6} = k\frac{r\sin\varphi + d_{26}}{r^2},$$

$$a_{7,7} = \frac{\cos\varphi}{r}, \quad a_{7,8} = -k\frac{d_{26}}{r}, \quad a_{7,9} = d_{21}\cos\varphi,$$

$$a_{7,10} = d_{21}\sin\varphi, \quad a_{7,12} = d_{23},$$

$$a_{7,13} = \frac{d_{24}}{r} + k^2\frac{d_{25}\sin\varphi}{r^2}, \quad a_{7,14} = -k^2\frac{d_{25}\cos\varphi}{r^2},$$

$$a_{7,15} = -k\frac{d_{24}r + d_{25}\sin\varphi}{r^2}, \quad a_{7,16} = \frac{d_{25}\cos\varphi}{r},$$

$$a_{7,1} = a_{7,2} = a_{7,4} = a_{7,11} = 0, a_{8,1} = d_{31} \cos \varphi, \quad a_{8,2} = d_{31} \sin \varphi,$$
$$a_{8,4} = d_{33}, a_{8,5} = \frac{d_{34}}{r} + k^2 \frac{d_{35} \sin \varphi}{r^2}, \quad a_{8,6} = -k^2 \frac{d_{35} \cos \varphi}{r^2},$$
$$a_{8,7} = k \frac{d_{34} r + d_{35} \sin \varphi}{r^2}, a_{8,8} = \frac{d_{35} \cos \varphi}{r}, \quad a_{8,11} = d_{32}, \quad a_{8,14} = -k \frac{d_{36}}{r^2},$$
$$a_{8,16} = k \frac{d_{36}}{r}, a_{8,3} = a_{8,9} = a_{8,10} = a_{8,12} = a_{8,13} = a_{8,15} = 0,$$

with:

$$f_1 = -q_r, \quad f_2 = -q_z, \quad f_3 = -q_\theta, \quad f_9 = -q'_x, \quad f_{10} = -q'_z, \quad f_{11} = -q'_\theta,$$
$$f_4 = f_5 = f_6 = f_7 = f_8 = 0, \quad f_{12} = f_{13} = f_{14} = f_{15} = f_{16} = 0. \tag{3.3.28}$$

For isotropic and anisotropic shells whose principal directions of elasticity coincide with directions of coordinate lines, we have for two matrices $A_2 = A_3 = 0$. In the case of orthotropic shells, the elements of the matrices A_1 and A_2 and the components of the vector $\bar{f}$ can be deduced from (3.3.27) and (3.3.28):

$$d_{11} = \frac{D_{11}}{\Delta}, \quad d_{12} = 0, \quad d_{13} = -\frac{K_{11}}{\Delta}, \quad d_{14} = \frac{K_{11}K_{12} - C_{12}D_{11}}{\Delta},$$
$$d_{15} = \frac{K_{11}D_{12} - D_{11}K_{12}}{\Delta}, \quad d_{16} = 0, \quad d_{17} = d_{11}, \quad d_{18} = d_{13}, \quad d_{19} = 0,$$
$$d_{21} = d_{23} = d_{24} = d_{25} = d_{27} = d_{28} = 0, \quad d_{22} = \frac{r^2}{\Delta_1},$$
$$\frac{v}{R} \frac{\cos \varphi}{R \sin \varphi} u_r - \sin \varphi \vartheta_s, \quad d_{29} = 0,$$
$$d_{31} = d_{13}, \quad d_{32} = 0, \quad d_{33} = \frac{C_{11}}{\Delta}, \quad d_{34} = \frac{K_{11}C_{12} - K_{12}C_{11}}{\Delta},$$
$$d_{35} = \frac{K_{11}K_{12} - C_{11}D_{12}}{\Delta}, \quad d_{36} = 0, \quad d_{37} = d_{13}, \quad d_{38} = d_{33}, \quad d_{39} = 0,$$
$$d_{41} = -d_{14}, \quad d_{42} = 0, \quad d_{43} = -d_{34}, \quad d_{44} = C_{22} + C_{12}d_{14} + K_{12}d_{34},$$
$$d_{45} = K_{22} + C_{12}d_{15} + K_{12}d_{35}, \quad d_{46} = 0, \quad d_{47} = -d_{14},$$
$$d_{48} = -d_{34}, \quad d_{49} = 0, \quad d_{51} = -d_{15}, \quad d_{52} = 0, \quad d_{53} = -d_{35},$$
$$d_{54} = d_{45}, \quad d_{59} = 0, \quad d_{55} = D_{22} + K_{12}d_{15} + D_{12}d_{35}, \quad d_{56} = 0,$$
$$d_{57} = -d_{15}, \quad d_{58} = -d_{35}, \quad d_{61} = d_{63} = d_{64} = d_{65} = d_{67} = d_{68} = 0,$$
$$\frac{v}{R} \frac{\cos \varphi}{R \sin \varphi} u_r - \sin \varphi \vartheta_s,$$
$$d_{66} = \frac{2r^2}{\Delta_1}\left(D_{66}C_{66} - K_{66}^2\right), \quad d_{69} = 0, \quad \Delta = C_{11}D_{11} - K_{11}^2,$$
$$\Delta_1 = C_{66}r^2 + 4 \sin \varphi (K_{66}r + D_{66} \sin \varphi). \tag{3.3.29}$$

For the case of orthotropic shells the functions B_{mp} that appear in the coefficients C_{mp}, K_{mp}, D_{mp} are expressed in terms of material constants of the ith layer as follows:

$$B_{11}^{i} = \frac{E_{s}^{i}}{1 - v_{s}^{i} v_{\theta}^{i}}, \quad B_{12}^{i} = \frac{v_{\theta}^{i} E_{s}^{i}}{1 - v_{s}^{i} v_{\theta}^{i}} = \frac{v_{s}^{i} E_{\theta}^{i}}{1 - v_{s}^{i} v_{\theta}^{i}},$$
$$B_{22}^{i} = \frac{E_{\theta}^{i}}{1 - v_{s}^{i} v_{\theta}^{i}}, \quad B_{66}^{i} = G_{s\theta}^{i}, \tag{3.3.30}$$

The stiffness characteristics appearing in (3.3.29) are determined by the expressions (1.4.12). For homogeneous isotropic shells it is necessary to put:

$$d_{11} = d_{17} = \frac{1}{D_N}, \quad d_{14} = -v, \quad d_{12} = d_{13} = d_{15} = d_{16} = d_{18} = d_{19} = 0,$$
$$d_{22} = \frac{2}{p_0(1 - v)D_N}, \quad d_{26} = -\frac{4 \sin \varphi D_M}{p_0 r D_N}, \tag{3.3.31}$$

$$d_{21} = d_{23} = d_{24} = d_{25} = d_{27} = d_{28} = d_{29} = 0, \quad d_{33} = d_{38} = \frac{1}{D_M}, \quad d_{35} = -v,$$
$$d_{31} = d_{32} = d_{34} = d_{36} = d_{37} = 0, \quad d_{39} = 0, \quad d_{41} = d_{47} = v,$$
$$d_{44} = (1 - v^2)D_N, \quad d_{42} = d_{43} = d_{45} = d_{46} = d_{48} = d_{49} = 0, d_{53} = d_{58} = v,$$
$$d_{55} = (1 - v^2)D_M, \quad d_{51} = d_{52} = d_{54} - d_{56} = d_{57} = d_{59} = 0,$$
$$d_{62} = \frac{2 \sin \varphi D_M}{p_0 r D_N}, \quad d_{66} = \frac{(1 - v)D_M}{p_0},$$
$$d_{61} = d_{63} = d_{64} = d_{65} = d_{67} = d_{68} = d_{69} = 0, \quad p_0 = 1 + \frac{4D_M \sin^2 \varphi}{D_N r^2}.$$

For isotropic and orthotropic shells the problem is reduced to the solution of two systems of eighth-order equations. These systems of equations should be solved simultaneously, when boundary conditions relate values with and without primes, i.e., when the force or the displacement, directed at an oblique angle to the plane of the meridional section and does not lie within it, are specified at the boundary contour.

The elements of the matrix A and components of the vector $\bar{f}$ do not include derivatives of the stiffness characteristics of the shells of revolution and therefore are of the same shape for the shells with constant and variable stiffness. The equations obtained also do not include the shell curvature $1/R_s$ in the meridional direction.

3.3.2 Stress State of a High-Pressure Balloon Made of a Glass-Reinforced Plastic

We will give an example of application of the method of discrete orthogonalization for the solution of the stress-strain state problem of shells of revolution—a high-pressure balloon. Strong and light high-pressure balloons made by winding are widely used in different fields of engineering (Fig. 3.6). Along with small weight and high specific strength, shatterproof fracture is a typical feature of these balloons.

The nearly ellipsoidal shape of the balloon results from the requirement of small circumferential stresses in the equatorial zone. It is guaranteed by a specific orientation of the fiberglass during winding, namely almost along the meridian.

A balloon with a capacity of 15 L was made of OSP-10E glass-reinforced plastic composed of alkali-free IS 55/6-250 fiberglass and EFB-4 epoxy-fenolic bonding.

The cross-section of the balloon is shown in Fig. 3.6, where "1" refers to the balloon casing, "2" is the connecting pipe, and "3" is the rubber seal. The angle of laying the fiberglasses relative to the meredian depends on the radius of the connecting pipes.

The balloons made by winding are of variable thickness along the meridian, which is maximum near the connecting pipe and minimum at the equator. The strain state of the balloon was analyzed based on the theory of orthotropic shells with variable stiffness. The balloon subjected to an internal pressure q_0 is a shell of revolution whose inner surface is formed by revolution of the curve

$$\frac{\mathrm{d}z}{\mathrm{d}r} = \frac{(r - r_N)^2}{\sqrt{a^4 - (r - r_N)^2}} \tag{3.3.32}$$

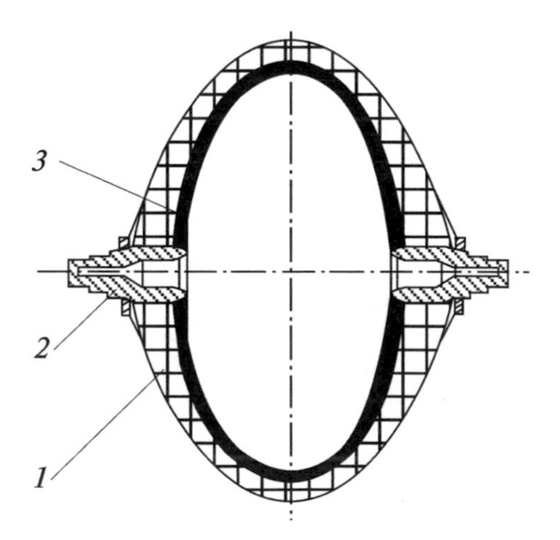

Fig. 3.6 High-pressure balloon made of fiberglass

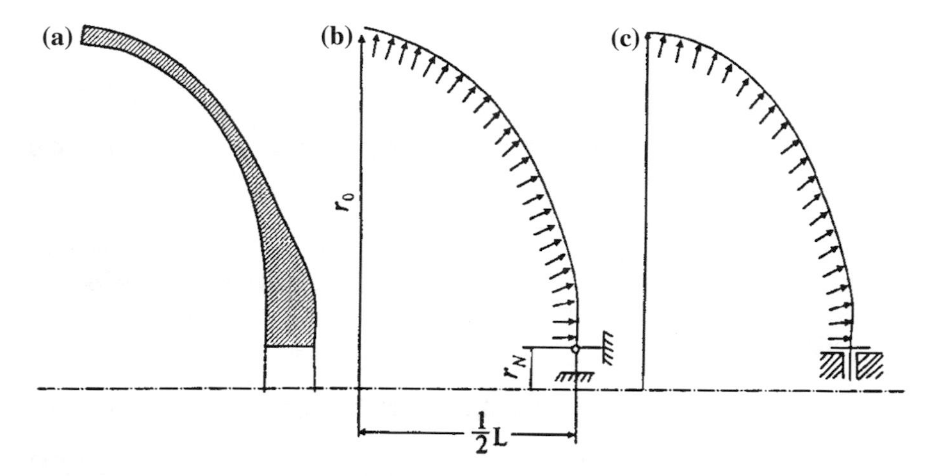

Fig. 3.7 Crossectional cuts of the balloon

about the Oz-axis, where 2 a is the spacing between the connecting pipes along the axis of revolution, r_N is the radius of a parallel contour near the connecting pipe. The meridional section of the balloon is shown in Fig. 3.7a.

Since the stress-strain state of the balloon is symmetrical relative to the plane of the cross-section at $\varphi = \pi/2$, it will suffice to consider only one half of the balloon, for example, the right one, after specifying symmetry conditions on the left hand side contour ("equator"), where $s = 0$:

$$N_x = \hat{S} = u_z = \vartheta_s = 0 \ \text{ at } \ s = 0. \tag{3.3.33}$$

The balloon thickness and the median surface are specified in the form of plots or tables. Hence, in order to describe the geometry of the balloon, the shell meridian on the interval from the "equator" to the boundary contour was broken into five segments in units of centimeters. These segments were approximated to sufficient accuracy by a circumference and four sections of straight lines specified by the following expressions: at the segment I (spherical shell) $r = 10.2 + 8.85\cos\frac{s}{8.85}$; $\sin\varphi = \cos\frac{s}{8.85}$, $\cos\varphi = -\sin\frac{s}{8.85}$, h = 0.5 + 0.0314386 s, $\gamma = 1$, $s_0 \le s \le s_1$; at the segment II (conical shell) r = 0.22793120 × 10 − 0.88112000 s, $\sin\varphi = 0.47289200$, $\cos\varphi = -0.88112000$, h = −0.326664 + 0.118069 s, $\gamma = 1$, $s_1 \le s \le s_2$; at the segment III (conical shell) $r = 0.23506987 \times 10^2 - 0.93632900\,s$, $\sin\varphi = 0.35112300$, $\cos\varphi = 0.93632900$, h = −0.817830 + 0.156055 s, $\gamma = 1$, $s_2 \le s \le s_3$; at the segment IV (conical shell) $r = 0.24020658 \times 10^2 - 0.96948300\,s$, $\sin\varphi = 0.24515700$, $\cos\varphi = -0.96948300$, h = −4.615455 + 0.401166 s, $\gamma = 1$, $s_3 \le s \le s_4$; at the segment V (plate) r = −s + 0.24630398 × 10^2, $\sin\varphi = 0$, $\cos\varphi = -1$, $h = 3,4$, $\gamma = 1, s_4 \le s \le s_N$.

For the calculations we used:

$$
\begin{aligned}
&s_0 = 0, s_1 = 0.95424240 \times 10, \quad s_2 = 0.12930271 \times 10^2,\\
&s_3 = 0.15493472 \times 10^2, \quad s_4 = 0.19980398 \times 10^2,\\
&s_N = 0.22330398 \times 10^2 (N = 5).
\end{aligned}
\tag{3.3.34}
$$

Since the boundary conditions at the junction of the balloon contour with the connecting pipe are not exactly determined, the calculations were carried out for two different variants of conditions at the contour—a hinge attachment and a rigid fixing (Fig. 3.7b, c)

$$
u_x = u_z = v = M_s = 0 \text{ at } s = s_N,
\tag{3.3.35}
$$

$$
u_x = u_z = v = \vartheta_s = 0 \text{ at } s = s_N.
\tag{3.3.36}
$$

The material of the shell is orthotropic where the principal directions of elasticity form some angle $\alpha = \alpha(s)$ with the meridian because the winding of the balloon is performed along geodesic lines. Because the winding when making the balloon is crossed, we write the elasticity relations for orthtropic shells with a variable along the meridian stiffness characteristics in the following form:

$$
E_i = E_i(s), \quad v_i = v_i(s), \quad G_{s\theta} = G_{s\theta}(s), \quad (i = s, \theta), \quad h = h(s).
\tag{3.3.37}
$$

In this case, the mechanical characteristics are determined as follows:

$$
\begin{aligned}
E_s &= \frac{E_1 E_2}{E_1 \frac{c^2}{r^2} + E_2\left(1 - \frac{c^2}{r^2}\right) + A\frac{c^2}{r^2}\left(1 - \frac{c^2}{r^2}\right)},\\
E_\theta &= \frac{E_1 E_2}{E_2 \frac{c^2}{r^2} + E_1\left(1 - \frac{c^2}{r^2}\right) + A\frac{c^2}{r^2}\left(1 - \frac{c^2}{r^2}\right)},\\
v_s &= \frac{E_s}{E_1 E_2}\left[v_1 E_2 + A\frac{c^2}{r^2}\left(1 - \frac{c^2}{r^2}\right)\right],\\
v_\theta &= \frac{E_\theta}{E_1 E_2}\left[v_1 E_2 + A\frac{c^2}{r^2}\left(1 - \frac{c^2}{r^2}\right)\right],\\
A &= E_1 E_2\left(\frac{1}{G} - \frac{2v_1}{E_1}\right) - E_1 - E_2.
\end{aligned}
\tag{3.3.38}
$$

Here E_1, E_2 are the elasticity moduli of the unidirectional glass-reinforced plastic in the longitudinal and transverse directions, respectively; v_1 is Poisson's ratio; G is the shear modulus, c is a constant, which depends on the winding angle α according to the equation $c/r = \sin\alpha$. For the zone of the connecting pipe we have $\alpha = \pi/4$, and the values v_1, G, as well as $E_s, E_\theta, v_s, v_\theta$ for some fixed winding angles were determined experimentally:

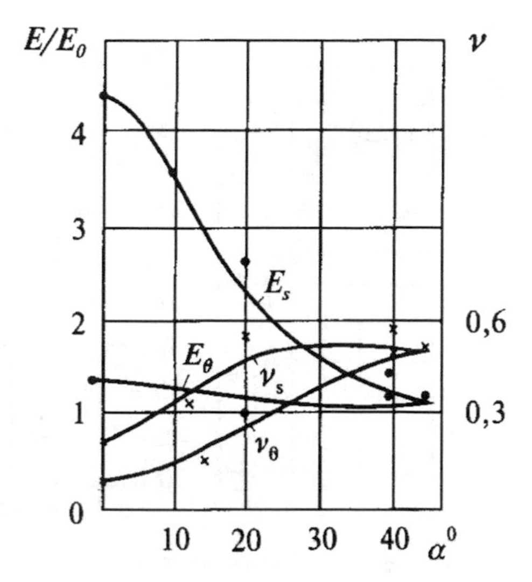

Fig. 3.8 Mechanical characteristics (see text)

$$\alpha = 0, \quad E_1 = 4.4E_0, \quad E_2 = 1.3E_0, \quad G = 0.37E_0, \\ \nu_1 = 0.18, E_0 = 0.981 \times 10^{10}\mathrm{N/m^2}. \tag{3.3.39}$$

The elastic characteristics of the material were determined by testing specimens cut by equivalent winding angles. The average values of mechanical characteristics are indicated in Fig. 3.8 by dots for E_s and E_θ and by crosses for ν_s and E_θ.

By taking the values of mechanical characteristics (3.3.39) for $\alpha = 0$ into account and by using the formulae (3.3.38), we find their distribution depending on the angle α (Fig. 3.8).

The plots reveal that the difference between the values of mechanical characteristics adopted in the calculations and those obtained experimentally is insignificant which supports the reliability of the data that was used. When solving the problem with the mechanical characteristics presented in Fig. 3.8, we found the values of all characteristics of the balloon stress-strain state. Figure 3.9 shows the distributions of the meridional and of the circumferential stresses, $\sigma_s^\pm$ and $\sigma_\theta^\pm$, respectively, on the outer and inner surfaces of the balloon. The solid line corresponds to the case when the connecting pipes are aligned and the dashed line to the case when the experimentally measured axial shifts of connecting pipes are taken into account. "I" refers to the stresses in the case of the hinged supporting of ends. "II" are the stresses in the case of rigid fixation. In the calculations, it was used that:

$$u_z = 1.75\,\mathrm{mm}(q_0 = 9.8 \times 10^6\ \mathrm{N/m^2}). \tag{3.3.40}$$

From the obtained results it follows that the influence of difference in the boundary conditions on the shell stress state at some distance from the place of

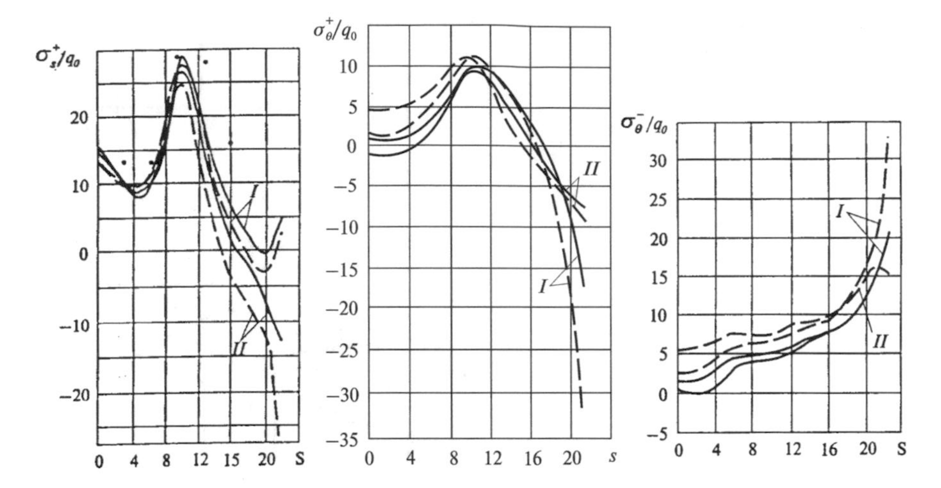

Fig. 3.9 Stress distributions (see text)

fixation is insignificant. This fact allows us to judge on the stress state of the balloon in the absence of any information about boundary conditions. It is also seen that the axial misalignment of the connecting pipes does not significantly influence the stress state of the balloon.

The maximum stresses in the balloon are the meridional stresses σ_s^+ in the vicinity of the point $s = 10$. The meridional stresses σ_s^- on inner surface are significantly smaller. The circumferential stresses $\sigma_\theta^\pm$ on both surfaces of the balloon are small compared with σ_s^+ due to the choice of the shape and way of the balloon fabrication.

In order to confirm the reliability of the calculation technique, we have evaluated the stress state of three balloons with capacity of 15 L by using strain gauges. The balloons were loaded by the internal pressure up to 100 atmospheres. As the result of tests, we determined average values of strains and associate values of meridional stresses $\sigma_s^\pm$ which are marked in the very left inset of Fig. 3.9 by points.

By comparing the experimental and the calculated data, we found that the best agreement is found at a distance from the connecting pipe in the case when the stresses are maximum. In the vicinity of the connecting pipe we observe some disagreement in the results, which is attributed to differences between the design schemes and the experiment. The satisfactory coincidence of the experimental data with results of calculations makes it possible to assume that the proposed technique for stress analysis of the balloons, based on the theory of anisotropic shells with variable stiffness, can be used for the design of balloons and similar items made of fiberglass promoting thereby the creation of light and strong structures.

3.4 Free Vibrations of Rectangular Plates

3.4.1 Governing Relations

Let us consider the problem on free vibrations of a rectangular orthotropic plate with the variable thickness h(x,y) in a rectangular coordinate system (the mid-surface Oxy of the plate is used as the coordinate plane). The problem is formulated within the framework of Kirchhoff-Love theory. In this case, the vibration equations can be written as:

$$\begin{aligned} &\frac{\partial Q_x}{\partial x} + \frac{\partial Q_y}{\partial y} = \rho h \frac{\partial^2 w}{\partial t^2}, \\ &\frac{\partial M_x}{\partial x} + \frac{\partial M_{xy}}{\partial y} = Q_x, \\ &\frac{\partial M_y}{\partial y} - \frac{\partial M_{xy}}{\partial x} = Q_y, \end{aligned} \quad (3.4.1)$$

where x, y are rectangular Cartesian coordinates $(0 \le x \le a, 0 \le y \le b)$, t is time, w is the deflection of the plate, and ρ is the mass density.

The following relations for the moments M_x, M_y, M_{xy} and shearing forces Q_x and Q_y hold:

$$\begin{aligned} M_x &= -\left(D_{11}\frac{\partial^2 w}{\partial x^2} + D_{12}\frac{\partial^2 w}{\partial y^2}\right), \\ Q_x &= -\left[D_{11}\frac{\partial^3 w}{\partial x^3} + (D_{12} + 2D_{66})\frac{\partial^3 w}{\partial x \partial y^2}\right], \\ M_y &= -\left(D_{12}\frac{\partial^2 w}{\partial x^2} + D_{22}\frac{\partial^2 w}{\partial y^2}\right), \\ Q_y &= -\left[D_{22}\frac{\partial^3 w}{\partial y^3} + (D_{12} + 2D_{66})\frac{\partial^3 w}{\partial x^2 \partial y}\right], \\ M_{xy} &= -2D_{66}\frac{\partial^2 w}{\partial x \partial y}, \end{aligned} \quad (3.4.2)$$

where the stiffness characteristics D_{ij} of the plate are determined by the formulae:

$$\begin{aligned} &D_{ij} = \frac{B_{ij} h^3(x,y)}{12}, \quad B_{11} = \frac{E_1}{1 - \nu_1 \nu_2}, \\ &B_{12} = \frac{\nu_2 E_1}{1 - \nu_1 \nu_2} = \frac{\nu_1 E_2}{1 - \nu_1 \nu_2}, \\ &B_{22} = \frac{E_2}{1 - \nu_1 \nu_2}, \quad B_{66} = G_{12}. \end{aligned}$$

Here $E_1, E_2, G_{12}, \nu_1, \nu_2$ are the elasticity modulus, shear modulus, and Poisson's ratios.

From the system of equations (3.4.1) and (3.4.2), we obtain the following equivalent system of differential equations with respect to the deflection w:

$$
\begin{aligned}
& D_{11}\frac{\partial^4 w}{\partial x^4} + D_{22}\frac{\partial^4 w}{\partial y^4} + 2(D_{12}+2D_{22})\frac{\partial^4 w}{\partial x^2 \partial y^2} + 2\frac{\partial D_{11}}{\partial x}\frac{\partial^3 w}{\partial x^3} + 2\frac{\partial D_{22}}{\partial y}\frac{\partial^3 w}{\partial y^3} \\
& \quad + 2\frac{\partial}{\partial y}(D_{12}+2D_{66})\frac{\partial^3 w}{\partial x^2 \partial y} + 2\frac{\partial}{\partial x}(D_{12}+2D_{66})\frac{\partial^3 w}{\partial x \partial y^2} \\
& \quad + \left(\frac{\partial^2 D_{11}}{\partial x^2} + \frac{\partial^2 D_{12}}{\partial y^2}\right)\frac{\partial^2 w}{\partial x^2} + \left(\frac{\partial^2 D_{12}}{\partial x^2} + \frac{\partial^2 D_{22}}{\partial y^2}\right)\frac{\partial^2 w}{\partial y^2} \\
& \quad + 4\frac{\partial^2 D_{66}}{\partial x \partial y}\frac{\partial^2 w}{\partial x \partial y} + \rho h\frac{\partial^2 w}{\partial t^2} = 0.
\end{aligned}
\tag{3.4.3}
$$

The boundary conditions, expressed in terms of deflection, are prescribed on the sides $x = 0$, $x = a$, $y = 0$, $y = b$. For $y = \text{const}$, the following boundary conditions are specified; the contours are rigidly fixed:

$$
w = 0, \quad \frac{\partial w}{\partial y} = 0 \ \text{ for } \ y = 0, \ \ y = b; \tag{3.4.4}
$$

the contours are hinged:

$$
w = 0, \quad \frac{\partial^2 w}{\partial y^2} = 0 \ \text{ for } \ y = 0, y = b; \tag{3.4.5}
$$

one contour is hinged, another is rigidly fixed:

$$
\begin{aligned}
w = 0, \quad \frac{\partial^2 w}{\partial y^2} = 0 \ \text{ for } \ y = 0, \\
w = 0, \quad \frac{\partial w}{\partial y} = 0 \ \text{ for } \ y = b.
\end{aligned}
\tag{3.4.6}
$$

Analogous conditions can be specified on the contours $x = \text{const}$.

3.4.2 Method of Solution

We will search for the solution of Eq. (3.4.3) in the form:

$$
w = \sum_{i=0}^{N} w_i(x)\psi_i(y), \tag{3.4.7}
$$

where $w_i(x)(i = \overline{1,N})$ are the unknown functions, $\psi_i(y)$ are functions generated with the help of the fifth-power B-splines ($N \geq 6$). The choice of the functions $\psi_i(y)$ is motivated by the necessity to satisfy boundary conditions at $y = \text{const}$ using linear combinations of B-splines:

$$\begin{aligned}
\psi_0(y) &= \alpha_{11}B_5^{-2}(y) + \alpha_{12}B_5^{-1}(y) + B_5^0(y),\\
\psi_1(y) &= \alpha_{21}B_5^{-1}(y) + \alpha_{22}B_5^0(y) + B_5^1(y),\\
\psi_2(y) &= \alpha_{31}B_5^{-2}(y) + \alpha_{32}B_5^0(y) + B_5^2(y),\\
\psi_i(y) &= B_5^i(y), \quad i = 3, 4, \ldots, N-3,\\
\psi_{N-2}(y) &= \beta_{31}B_5^{N+2}(y) + \beta_{32}B_5^N(y) + B_5^{N+2}(y),\\
\psi_{N-1}(y) &= \beta_{21}B_5^{N+1}(y) + \beta_{22}B_5^N(y) + B_5^{N-1}(y),\\
\psi_N(y) &= \beta_{11}B_5^{N+2}(y) + \beta_{12}B_5^{N+1}(y) + B_5^N(y),
\end{aligned} \tag{3.4.8}$$

where $B_5^i(y)$ (i = −2, …, N + 2, i is the spline number) are the splines constructed on the uniform mesh Δ with the step h_y:

$$y_{-5} < y_{-4} < \cdots < y_N < y_{N+5} < \cdots < y_{N+5}; \quad y_0 = 0; \quad y_N = b:$$

$$B_5^i(y) = \frac{1}{120} \times \begin{cases}
0, & \infty < y < y_{i-3},\\
z^5, & y_{i-3} \leq y < x_{i-2},\\
-5z^5 + 5z^4 + 10z^3 + 10z^2 + 5z + 1, & y_{i-2} \leq y < y_{i-1},\\
10z^5 - 20z^4 - 20z^3 + 20z^2 + 50z + 26, & y_{i-1} \leq y < y_i,\\
-10z^5 + 30z^4 - 60z^2 + 66, & y_i \leq y < y_{i+1},\\
5z^5 - 20z^4 + 20z^3 + 20z^2 - 50z + 26, & y_{i+1} \leq y < y_{i+2},\\
(1-z)^5, & y_{i+2} \leq y < y_{i+3},\\
0, & y_{i+3} \leq y < \infty,
\end{cases}$$

where $z = \frac{y - y_k}{h_y}$ on the interval $[y_k, y_{k+1}]$, $k = \overline{i-3, i+2}$; $i = \overline{-3, N+2}$; $h_y = y_{k+1} - y_k = \text{const}$; α_{ij} and $\beta_{ij}(i = 1, 2, 3; j = 1, 2)$ are the constant coefficients, which are determined depending on the specified boundary conditions on the plate edges $y = 0$ and $y = b$, respectively.

Let us designate two matrices by $A_\alpha = \begin{bmatrix} \alpha_{11} & \alpha_{12} \\ \alpha_{21} & \alpha_{22} \\ \alpha_{31} & \alpha_{32} \end{bmatrix}$, and $A_\beta = \begin{bmatrix} \beta_{11} & \beta_{12} \\ \beta_{21} & \beta_{22} \\ \beta_{31} & \beta_{32} \end{bmatrix}$.

Then, under conditions of rigid fixation of edges $y = 0$ and $y = b$ we have:

$$A_\alpha = A_\beta = \begin{bmatrix} \frac{165}{4} & -\frac{33}{8} \\ 1 & -\frac{26}{33} \\ 1 & -\frac{1}{33} \end{bmatrix},$$

and under conditions of hinged edges $y = 0$, $y = b$:

$$A_{\alpha} = A_{\beta} = \begin{bmatrix} 12 & -3 \\ -1 & 0 \\ -1 & 0 \end{bmatrix},$$

and under conditions (3.4.4) on the plate edges:

$$A_{\alpha} = \begin{bmatrix} \frac{165}{4} & -\frac{33}{8} \\ 1 & -\frac{26}{33} \\ 1 & -\frac{1}{33} \end{bmatrix}, \quad A_{\beta} = \begin{bmatrix} 12 & -3 \\ -1 & 0 \\ -1 & 0 \end{bmatrix}.$$

Let us write Eq. (3.4.3) in the form

$$\begin{aligned} \frac{\partial^4 w}{\partial x^4} = a_1 \frac{\partial^3 w}{\partial x^3} + a_2 \frac{\partial^4 w}{\partial x^2 \partial y^2} + a_3 \frac{\partial^3 w}{\partial x^2 \partial y} + a_4 \frac{\partial^2 w}{\partial x^2} \\ + a_5 \frac{\partial^3 w}{\partial x \partial y^2} + a_6 \frac{\partial^2 w}{\partial x \partial y} + a_7 \frac{\partial^4 w}{\partial y^4} + a_8 \frac{\partial^3 w}{\partial y^3} + a_9 \frac{\partial^2 w}{\partial y^2} + a_{10} w, \end{aligned} \tag{3.4.9}$$

where $a_i = a_i(x, y), i = 1, 2, \ldots, 9, a_{10} = a_{10}(x, y, \omega)$. By substituting expression (3.4.7) into Eq. (3.4.9), we will require it to be satisfied at the prescribed collocation points $\xi_k \in [0, b], k = 0, N$.

Let us consider the case, where the number of mesh nodes is even, i.e., $N = 2n + 1 (n \geq 3)$, and collocation nodes satisfy conditions $\xi_{2i} \in [y_{2i}, y_{2i+1}]$, $\xi_{2i+1} \in [y_{2i}, y_{2i+1}]$, $(i = 0, 1, \ldots, n)$. Then we have two collocation nodes on the segment $[y_{2i}, y_{2i+1}]$ and no collocation nodes on the adjacent segments $[y_{2i+1}, y_{2i+2}]$. The collocation points on each segment are chosen in the following way: $\xi_{2i} = x_{2i} + z_1 h_y$, $\xi_{2i+1} = y_{2i} + z_2 h_y$ $(i = 0, 1, 2, \ldots, n)$, where z_1 and z_2 are the roots of the Legandre second-order polynomial on the segment [0,1]: $z_1 = \frac{1}{2} - \frac{\sqrt{3}}{6}$, $z_2 = \frac{1}{2} + \frac{\sqrt{3}}{6}$. Such a choice of collocation points is optimal and essentially increases the order of the approximation accuracy. As a result, we obtain a system of N + 1 linear differential equations with respect to w_i. By introducing the designations

$$\begin{aligned} &\Psi_j = [\psi_i^{(j)}(\xi_k)], \quad k, i = 0, \ldots, N, \quad j = 0, \ldots, 4, \\ &\bar{w} = \{w_0, w_1, \ldots, w_N\}^T, \\ &\bar{a}_r^T = \{a_r(x, \xi_0), a_r(x, \xi_1), \ldots, a_r(x, \xi_N)\} \quad , r = 1, \ldots, 9, \\ &\bar{a}_{10}^T = \{a_{10}(x, \xi_0, \omega), \quad a_r(x, \xi_1, \omega), \ldots, a_r(x, \xi_N, \omega)\}, \end{aligned}$$

$A = [a_{ij}]\ \{i,j = 0,\ldots,N)$ and denoting the matrix $[c_i a_{ij}]$ by $\bar{c} * A$, the system of differential equations will become:

$$\bar{w}^{\mathrm{IV}} = \Psi_0^{-1}(\bar{a}_7 * \Psi_4 + \bar{a}_8 * \Psi_3 + \bar{a}_9 * \Psi_2 + \bar{a}_{10} * \Psi)\bar{w} + \Psi_0^{-1}(\bar{a}_5 * \Psi_2 + \bar{a}_6 * \Psi_1)\bar{w}' + \Psi_0^{-1}(\bar{a}_2 * \Psi_3 + \bar{a}_3 * \Psi_1 \bar{a}_4 * \Psi_0)\bar{w}'' + \Psi_0^{-1}(\bar{a}_1 * \Psi_0)\bar{w}'''. \tag{3.4.10}$$

This system may be reduced to the normal form:

$$\frac{\mathrm{d}\bar{Y}}{\mathrm{d}x} = A(x,\omega)\bar{Y} \quad (0 \le x \le a), \tag{3.4.11}$$

where:

$$\bar{Y} = \{w_1, w_2, \ldots, w_{N+1}, w_1', w_2', \ldots, w_{N+1}', w_1'', w_2'', \ldots, w_{N+1}'', w_1''', w_2''', \ldots, w_{N+1}'''\}^T;$$
$$w_K^{(I)} = w^{(I)}(x, \xi_K), \quad K = 1,\ldots,N+1; \quad I = 1,2,3,$$

and $A(x,\omega)$ is a square matrix of the $(N+1) \times (N+1)$ th order. Boundary conditions for the given system can be written as:

$$B_1\bar{Y}(0) = \bar{b}_1, \quad B_2\bar{Y}(a) = \bar{b}_2, \tag{3.4.12}$$

where $\bar{b}_1$ and $\bar{b}_2$ are the zero vectors.

The eigenvalue problem for the system of ordinary differential equations (3.4.11) in combination with the boundary conditions (3.4.12) was solved by the method of discrete orthogonalization in combination with the step-by-step method (Grigorenko and Tregubenko [5]).

Based on the proposed technique we have studied the spectrum of natural vibrations of a square variable-thickness plate under different boundary conditions on edges. The plate thickness varied by the law $h(x) = [\alpha(6x^2 - 6x + 1) + 1]h_0$.

The plate has the shapes of cross-section shown in Fig. 3.10. It is fabricated of orthotropic glass-reinforced plastic with the elasticity moduli $E_1 = 4.76 \times 10^4$ MPa and $E_2 = 2.07 \times 10^4$ MPa, shear modulus $G_{12} = 0.531 \times 10^4$ MPa, and Poisson's ratios $v_1 = 0.149$ and $v_1 = 0.0647$.

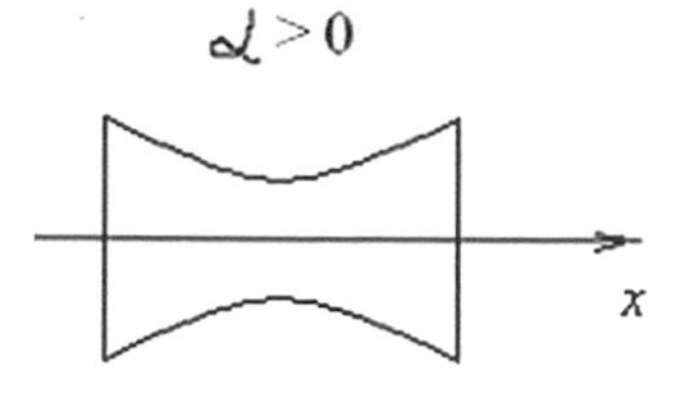

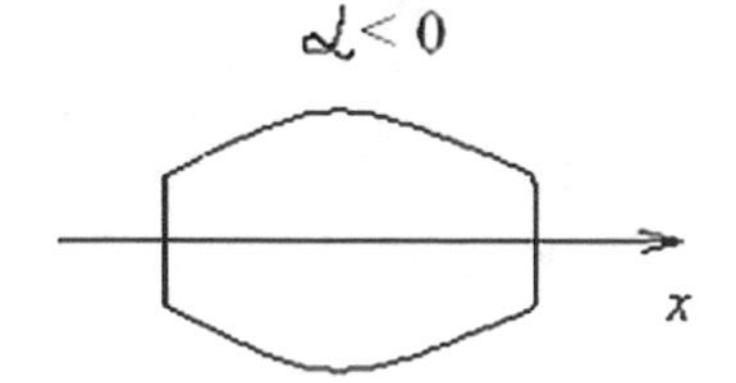

Fig. 3.10 Cross-sectional shapes

Table 3.11 Spline-collocation method for a clamped plate

ω	N						
	10	12	14	16	18	20	22
$\alpha = 0$							
ω_1	61.139	61.132	61.129	61.127	61.127	61.127	61.127
ω_2	107.188	107.066	107.016	106.994	106.982	106.976	106.972
ω_3	142.550	142.537	142.532	142.530	142.529	142.528	142.528
$\alpha = 0.3$							
ω_1	62.108	62.102	62.099	62.099	62.098	62.098	62.098
ω_2	97.737	97.637	97.598	97.580	97.570	97.566	97.562
ω_3	145.289	145.276	45.271	145.269	145.268	145.268	145.267

The dimensionless frequencies $\bar{\omega} = \omega a^2(\rho h/D_0)^{1/2}, D_0 = \frac{1}{12}h_0^3 \times 10^4$ MPa obtained by the spline-collocation method for a clamped plate and different number of collocation points (N = 10, 12, 14, 16, 18, 20, 22) differ slightly (see Table 3.11).

Tables 3.12 and 3.13 summarize dimensionless frequencies $\bar{\omega}_i$ $(i = 1, 2, 3)$ (ordered by value) for the orthotropic square plate for $\alpha \leq 0$ and $\alpha > 0$, respectively;

1. all edges are clamped, boundary conditions of the type A:

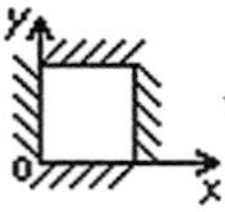

   $w = 0, \frac{\partial w}{\partial y} = 0$ for $y = 0, y = a$, and $w = 0, \frac{\partial w}{\partial x} = 0$ for $x = 0$, $x = a$;

2. three edges are clamped and the fourth one is hinged, boundary conditions of the type B:

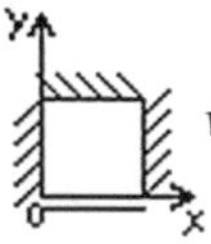

   $w = 0, \frac{\partial w}{\partial y} = 0$ for $y = a, w = 0, \frac{\partial w}{\partial x} = 0$ for $x = 0, x = a$, and $w = 0$, $\frac{\partial^2 w}{\partial y^2} = 0$ for $y = 0$;and boundary conditions of the type C:

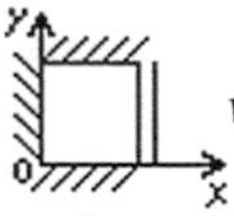

   $w = 0, \frac{\partial w}{\partial y} = 0$ for $y = 0, y = a, w = 0, \frac{\partial w}{\partial x} = 0$ for $x = 0$, and $w = 0, \frac{\partial^2 w}{\partial x^2} = 0$ for $x = a$;

3. two edges are clamped and two are hinged, boundary conditions of the type D:

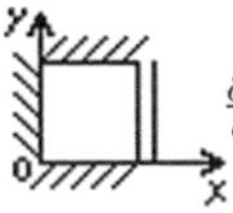

   $\frac{\partial w}{\partial y} = 0$ for $y = 0, w = 0, \frac{\partial^2 w}{\partial y^2} = 0$ for $y = a, w = 0, \frac{\partial w}{\partial x} = 0$ for $x = 0$, and $w = 0, \frac{\partial^2 w}{\partial x^2} = 0$ for $x = a$;

   boundary conditions of the type E:

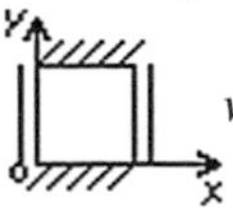

   $w = 0, \frac{\partial w}{\partial y} = 0$ for $y = 0, y = a, w = 0, \frac{\partial^2 w}{\partial x^2} = 0$ for $x = 0, x = a$;

Table 3.12 Dimensionless frequencies

Boundary conditions	ω	α					
		−0.5	−0.4	−0.3	−0.2	−0.1	0
A	ω_1	58.375	59.012	59.605	60.159	60.674	61.139
	ω_2	121.526	119.010	116.240	113.311	110.281	107.188
	ω_3	124.339	129.656	134.030	137.585	140.405	142.550
B	ω_1	50.227	51.605	52.905	54.29	54.270	56.320
	ω_2	102.334	100.723	98.953	97.083	95.147	93.166
	ω_3	121.396	126.863	131.384	135.083	138.045	140.330
C	ω_1	52.733	52.211	51.624	50.995	50.337	49.659
	ω_2	109.151	112.403	110.490	107.334	104.080	100.783
	ω_3	116.222	113.496	114.497	116.613	117.777	118.403
D	ω_1	43.939	44.009	43.998	43.923	43.790	43.607
	ω_2	97.299	95.178	92.905	90.549	88.155	85.755
	ω_3	105. 958	109.322	111.879	113.750	115.019	115.748
E	ω_1	48.701	47.412	46.059	44.676	43.289	41.918
	ω_2	95.664	97.177	97.994	98.235	97.985	96.671
	ω_3	113.313	110.306	107.049	103.676	100.170	97.306
G	ω_1	45.045	46.929	48.698	50.354	51.893	53.306
	ω_2	85.738	85.046	84.248	83.378	82.453	81.479
	ω_3	119.425	124.989	129.605	133.396	136.448	138.819

Table 3.13 Dimensionless frequencies

Boundary conditions	ω	α				
		0.1	0.2	0.3	0.4	0.5
A	ω_1	61.542	61.871	62.108	62.237	62.238
	ω_2	104.058	100.904	97.737	94.557	91.364
	ω_3	144.062	144.968	145.289	145.035	144.212
B	ω_1	57.266	58.096	58.791	59.339	59.724
	ω_2	91.149	89.104	87.026	84.911	82.749
	ω_3	144.977	143.017	143.469	138.723	131.720
C	ω_1	48.965	48.257	47.535	46.801	46.051
	ω_2	97.477	94.187	90.928	87.713	84.546
	ω_3	118.535	118.208	117.477	116.269	114.688
D	ω_1	43.376	43.098	42.744	42.402	41.982
	ω_2	83.369	81.013	78.694	76.420	74.190
	ω_3	115.981	115.752	115.085	113.998	112.506
E	ω_1	40.583	39.297	38.077	36.933	35.878
	ω_2	93.185	89.743	86.364	38.065	79.858
	ω_3	96.249	94.852	93.149	91.160	88.934
G	ω_1	54.586	55.718	56.642	57.495	58.109
	ω_2	80.460	79.389	78.258	77.055	75.765
	ω_3	140.291	34.815	29.283	23.717	18.135

boundary conditions of the type G:

$$w = 0, \frac{\partial^2 w}{\partial y^2} = 0 \text{ for } y = 0, y = a, w = 0, \frac{\partial w}{\partial x} = 0 \text{ for } x = 0, x = a.$$

The number of collocation points was N = 10. Dependencies of the dimensionless vibration frequency $\bar{\omega}_i$ of the orthotropic plate under different boundary conditions on the value of the parameter α are presented in Fig. 3.11. Maxima and minima at the plots of frequencies $\bar{\omega}_2$ and $\bar{\omega}_3$ versus α correspond to the mode restructuring.

The frequency of the clamped plate is the highest one under all considered boundary conditions on plate edges and different values of α. The first frequency for the boundary conditions of the type D slightly varies compared with other boundary conditions. Figure 3.12 shows modes of natural vibrations of the plate with boundary conditions of the type G.

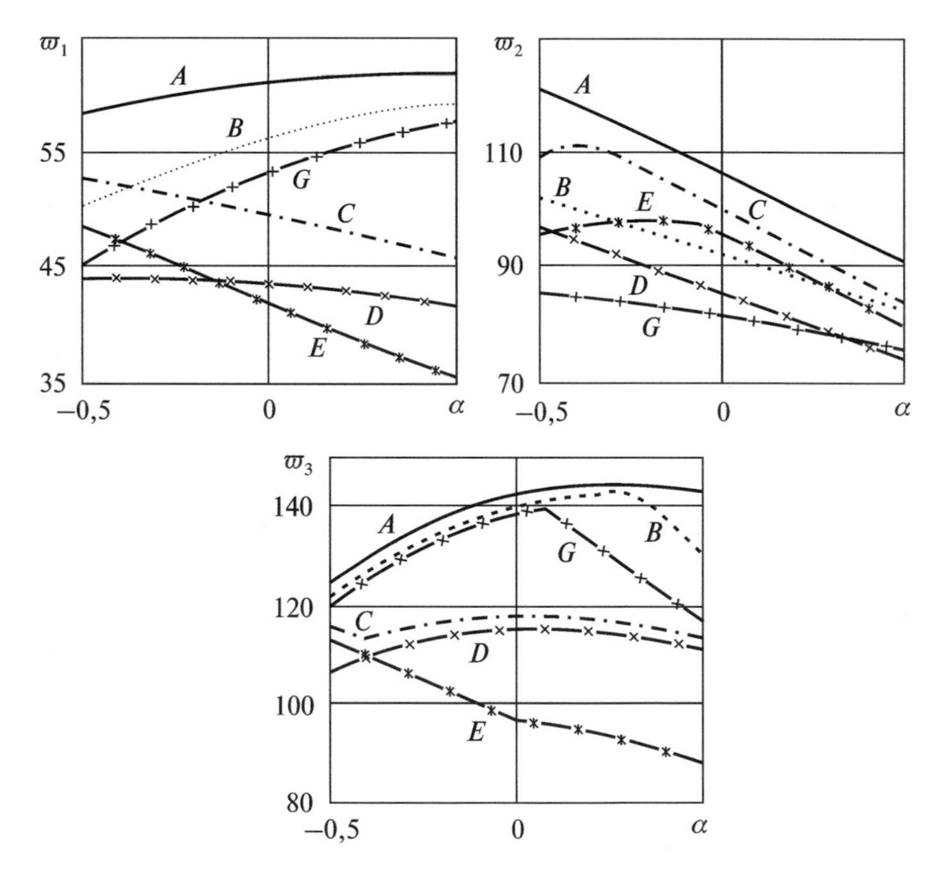

Fig. 3.11 Dimensionless vibration frequencies

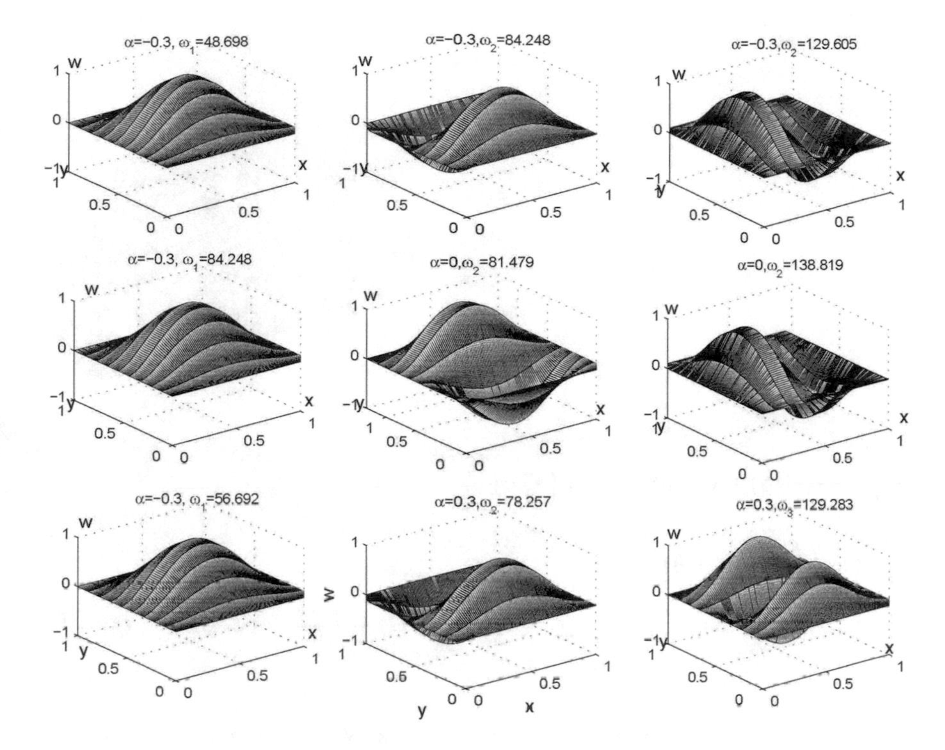

Fig. 3.12 Modes of natural vibrations

3.5 Free Vibrations of Conical Shells

3.5.1 Problem Formulation

Consider a freely vibrating isotropic conical shell of variable thickness $h(x, y)$ described in a curvilinear orthogonal coordinate system (s, θ), where s is the meridional coordinate; θ is the azimuth angle (Fig. 3.13). The Lamé parameters are given by $A = 1, B = r$, and the radii of principal curvatures $R_s = 0; R_\theta = r/\sin\varphi$, where r is the radius of a parallel circle; φ is the angle between the normal to the mid-surface and the axis of revolution.

The Kirchhoff-Love theory of thin shells employs the following equations to describe the free vibrations of conical shells:

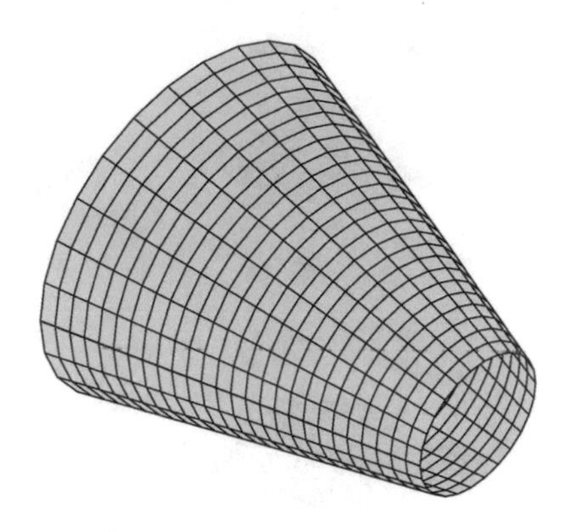

Fig. 3.13 Conical shell

$$\frac{\partial N_\theta}{\partial \theta}+\frac{\partial}{\partial s}(rS)+\cos\varphi S+\sin\varphi\left(Q_\theta+\frac{\partial H}{\partial s}\right)=r\rho h\frac{\partial^2 v(s,\theta,t)}{\partial t^2},$$
$$\frac{\partial}{\partial s}(rQ_s)+\frac{\partial Q_\theta}{\partial \theta}-\sin\varphi N_\theta=r\rho h\frac{\partial^2 w(s,\theta,t)}{\partial t^2},$$
$$\frac{\partial}{\partial s}(rM_s)+\frac{\partial H}{\partial \theta}-\cos\varphi M_\theta-rQ_s=0;\quad \frac{\partial}{\partial s}(rH)+\frac{\partial M_\theta}{\partial \theta}+\cos\varphi H-rQ_\theta=0,\tag{3.5.1}$$

where s,θ are curvilinear orthogonal coordinates of a point on the midsurface $(s_0\le s\le s_a, 0\le\theta\le b)$, t is time, u, v, w are the displacements of the mid-surface, and ρ is the density of the material. Strains and displacements are related by:

$$\begin{aligned}\varepsilon_s&=\frac{\partial u}{\partial s};\quad \varepsilon_\theta=\frac{1}{r}\frac{\partial v}{\partial \theta}+\frac{\cos\varphi}{r}u+\frac{\sin\varphi}{r}w,\\ \varepsilon_{s\theta}&=\frac{1}{r}\frac{\partial u}{\partial \theta}+\frac{\partial v}{\partial s}-\frac{\cos\varphi}{r}v,\quad \chi_s=-\frac{\partial^2 w}{\partial s^2},\\ \chi_\theta&=-\frac{1}{r^2}\frac{\partial^2 w}{\partial \theta^2}-\frac{\cos\varphi}{r}\frac{\partial w}{\partial s},\quad \chi_{s\theta}=\frac{\cos\varphi}{r^2}\frac{\partial w}{\partial \theta}-\frac{1}{r}\frac{\partial^2 w}{\partial s\partial \theta}.\end{aligned}\tag{3.5.2}$$

If the material of the shells is isotropic, the normal and shearing forces N_s, N_θ, S, bending and twisting moments M_s, M_θ, and H are expressed in terms of strains by:

$$\begin{aligned}&N_s=D_N(\varepsilon_s+v\varepsilon_\theta),\quad N_\theta=D_N(v\varepsilon_s+\varepsilon_\theta),\quad S=\frac{1-v}{2}D_N\varepsilon_{s\theta},\\ &M_s=D_M(\chi_s+v\chi_\theta),\quad M_\theta=D_M(v\chi_s+\chi_\theta),\quad H=D_M(1-v)\chi_{s\theta}.\end{aligned}\tag{3.5.3}$$

The stiffnesses of the shell are given by $D_N=Eh/(1-v^2)$, $D_M=Eh^3/12(1-v^2)$, where E and v are the elastic modulus and Poisson's ratio of the shell material, respectively.

The mid-surface undergoes small vibrations, and the displacements $u(s,\theta,t), v(s,\theta,t), w(s,\theta,t)$ can be represented by harmonic functions:

$$u(s,\theta,t) = u(s,\theta)\mathrm{e}^{-\mathrm{i}\omega t},\quad v(s,\theta,t) = v(s,\theta)\mathrm{e}^{-\mathrm{i}\omega t},\quad w(s,\theta,t) = w(s,\theta)\mathrm{e}^{-\mathrm{i}\omega t}, \tag{3.5.4}$$

where ω is the natural frequency.

From the system of equations (3.5.1)–(3.5.3) with (3.5.4), we derive three equivalent differential equations for the displacements u, v, w:

$$\begin{aligned}
\frac{\partial^2 u}{\partial\theta^2} &= F_u\left(u,\frac{\partial u}{\partial s},\frac{\partial^2 u}{\partial s^2},\frac{\partial u}{\partial\theta},v,\frac{\partial v}{\partial s},\frac{\partial v}{\partial\theta},\frac{\partial^2 v}{\partial s\partial\theta},w,\frac{\partial w}{\partial s},\omega\right),\\
\frac{\partial^2 v}{\partial\theta^2} &= F_v\left(u,\frac{\partial u}{\partial s},\frac{\partial u}{\partial\theta},\frac{\partial^2 u}{\partial s\partial\theta},v,\frac{\partial v}{\partial s},\frac{\partial v}{\partial\theta},\frac{\partial^2 v}{\partial s^2},w,\frac{\partial w}{\partial s},\frac{\partial w}{\partial\theta},\frac{\partial^2 w}{\partial s^2},\frac{\partial^2 w}{\partial\theta^2},\frac{\partial^2 w}{\partial s\partial\theta},\frac{\partial^3 w}{\partial s^2\partial\theta},\frac{\partial^3 w}{\partial\theta^3},\omega\right)\\
\frac{\partial^4 w}{\partial\theta^4} &= F_w\left(u,\frac{\partial u}{\partial s},\frac{\partial v}{\partial\theta},w,\frac{\partial w}{\partial s},\frac{\partial w}{\partial\theta},\frac{\partial^2 w}{\partial s^2},\frac{\partial^2 w}{\partial\theta^2},\frac{\partial^2 w}{\partial s\partial\theta},\frac{\partial^3 w}{\partial s^3},\frac{\partial^3 w}{\partial\theta^3},\frac{\partial^3 w}{\partial s\partial\theta^2},\frac{\partial^3 w}{\partial s^2\partial\theta},\frac{\partial^4 w}{\partial s^4},\frac{\partial^4 w}{\partial s^2\partial\theta^2},\omega\right),
\end{aligned} \tag{3.5.5}$$

where F_u, F_v, F_w are linear differential operators.

Boundary conditions for displacements are prescribed on the edges $s = s_0, s_a$, and $\theta = 0, b$. The following boundary conditions can be specified on the edges $s = \mathrm{const}$:

(i) clamped edge:

$$u = v = w = \frac{\partial w}{\partial s} \quad \text{at } s = s_0, s = s_a; \tag{3.5.6}$$

(ii) hinged edge:

$$v = \frac{\partial u}{\partial s} = w = \frac{\partial^2 w}{\partial s^2} \quad \text{at } s = s_0, s = s_a. \tag{3.5.7}$$

Similar conditions can be specified on the edges $\theta = \mathrm{const}$:

(i) clamped edge:

$$u = v = w = \frac{\partial w}{\partial\theta} \quad \text{at } \theta = 0, \theta = b; \tag{3.5.8}$$

(ii) hinged edge:

$$u = \frac{\partial v}{\partial\theta} = w = \frac{\partial^2 w}{\partial\theta^2} \quad \text{at } \theta = 0, \theta = b. \tag{3.5.9}$$

3.5.2 Method of Solution

The solution of the system of equations (3.5.5) is represented in the form:

$$u=\sum_{i=0}^{N}u_i(\theta)\varphi_i(s);\quad v=\sum_{i=0}^{N}v_i(\theta)\chi_i(s);\quad w=\sum_{i=0}^{N}w_i(\theta)\psi_i(s), \tag{3.5.10}$$

where $u_i(\theta)$, $v_i(\theta)$, $w_i(\theta)$ $(i=0,\ldots,N)$ are unknown functions; $\varphi_i(s)$, $\chi_i(s)$ are functions set up by using cubic B-splines $(N\geq 4)$; $\psi_i(s)$ are functions constructed by using quintic B-splines $(N\geq 6)$. The functions $\varphi_i(s)$, $\chi_i(s)$, $\psi_i(s)$ are chosen such that the boundary conditions at $s=\text{const}$ are satisfied when using linear combinations of cubic and quintic B-splines, respectively. We rearrange the system of equations (3.5.5) by:

$$\begin{aligned}
\frac{\partial^2 u}{\partial\theta^2} &= a_{11}\frac{\partial u}{\partial\theta}+a_{12}\frac{\partial^2 u}{\partial s^2}+a_{13}\frac{\partial u}{\partial s}+a_{14}u+a_{15}\frac{\partial^2 v}{\partial s\partial\theta}+a_{16}\frac{\partial v}{\partial\theta}+a_{17}\frac{\partial v}{\partial s}\\
&\quad+a_{18}v+a_{19}\frac{\partial w}{\partial s}+a_{110}w+a_{111}(\omega)u,\\
\frac{\partial^2 v}{\partial\theta^2} &= a_{21}\frac{\partial^2 u}{\partial s\partial\theta}+a_{22}\frac{\partial u}{\partial\theta}+a_{23}\frac{\partial u}{\partial s}+a_{24}u+a_{25}\frac{\partial v}{\partial\theta}+a_{26}\frac{\partial^2 v}{\partial s^2}+a_{27}\frac{\partial v}{\partial s}\\
&\quad+a_{28}v+a_{29}\frac{\partial^3 w}{\partial\theta^3}+a_{210}\frac{\partial^2 w}{\partial\theta^2}+a_{211}\frac{\partial^3 w}{\partial s^2\partial\theta}+a_{212}\frac{\partial^2 w}{\partial s\partial\theta}+a_{213}\frac{\partial w}{\partial\theta}\\
&\quad+a_{214}\frac{\partial^2 w}{\partial s^2}+a_{215}\frac{\partial w}{\partial s}+a_{216}w+a_{217}(\omega)v,\\
\frac{\partial^4 w}{\partial\theta^4} &= a_{31}\frac{\partial u}{\partial s}+a_{32}u+a_{33}\frac{\partial v}{\partial\theta}+a_{34}\frac{\partial^3 w}{\partial\theta^3}+a_{35}\frac{\partial^4 w}{\partial s^2\partial\theta^2}+a_{36}\frac{\partial^3 w}{\partial s\partial\theta^2}\\
&\quad+a_{37}\frac{\partial^2 w}{\partial\theta^2}+a_{38}\frac{\partial^3 w}{\partial s^2\partial\theta}+a_{39}\frac{\partial^2 w}{\partial s\partial\theta}+a_{310}\frac{\partial w}{\partial\theta}+a_{311}\frac{\partial^4 w}{\partial s^4}+a_{312}\frac{\partial^3 w}{\partial s^3}\\
&\quad+a_{313}\frac{\partial^2 w}{\partial s^2}+a_{314}\frac{\partial w}{\partial s}+a_{315}w+a_{316}(\omega)w,
\end{aligned} \tag{3.5.11}$$

where $a_{1n}=a_{1n}(s,\theta)$, $a_{2m}=a_{2m}(s,\theta)$, $a_{3k}=a_{3k}(s,\theta)$; $n=1,\ldots,10$, $m=1,\ldots,16$, $k=1,\ldots,15$, $a_{111}=a_{111}(s,\theta,\omega)$, $a_{217}=a_{217}(s,\theta,\omega)$, $a_{316}=a_{316}(s,\theta,\omega)$.

When substituting (3.5.10) into Eq. (3.5.11), we require that they hold at given collocation points $\xi_k\in[s_a,s_b]$, $k=0,\ldots,N$. If the number of knots $(N=2n+1$, $n\geq 3)$ is even and $\xi_{2i}\in[s_{2i},s_{2i+1}]$, $\xi_{2i+1}\in[s_{2i},s_{2i+1}]$, $(i=0,\ldots,n)$, there are two collocation points within the interval $[s_{2i},s_{2i+1}]$ and no collocation points with the adjacent intervals $[s_{2i+1},s_{2i+2}]$. The collocation points within each interval $[s_{2i},s_{2i+1}]$ are chosen such that $\xi_{2i}=s_{2i}+z_1h$, $\xi_{2i+1}=s_{2i}+z_2h$, $(i=0,\ldots,n)$, where h is the mesh spacing; z_1 and z_2 are the roots of a Legendre quadratic on the interval [0, 1], $z_1 = 1/2 - \sqrt{3}/6$ and $z_2 = 1/2 - \sqrt{3}/6$. This choice of collocation points is optimal and significantly improves the accuracy of approximation. After all transformations, we obtain a system of $N+1$ linear differential equations for u_i, v_i, w_i. If:

$$
\begin{aligned}
\Phi_l &= [\varphi_i^{(l)}(\xi_k)], X_l = [\chi_i^{(l)}(\xi_k)], \Psi_m = [\psi_i^{(m)}(\xi_k)], \\
&\quad (i,k = 0,\ldots,N, \quad l = 0,\ldots,2, \quad m = 0,\ldots,4), \\
\bar{u}^T &= \{u_0,\ldots,u_N\}, \bar{v}^T = \{v_0,\ldots,v_N\}, \bar{w}^T = \{w_0,\ldots,w_N\}, \\
\bar{a}_{1r}^T &= \{a_{1r}(\theta,\xi_0),\ldots,a_{1r}(\theta,\xi_N)\} \ (r = 1,\ldots,10), \\
\bar{a}_{2r}^T &= \{a_{2r}(\theta,\xi_0),\ldots,a_{2r}(\theta,\xi_N)\} \ (r = 1,\ldots,16) \\
\bar{a}_{3r}^T &= \{a_{3r}(\theta,\xi_0),\ldots,a_{3r}(\theta,\xi_N)\} \ (r = 1,\ldots,15) \\
\bar{a}_{111}^T &= \{a_{111}(\theta,\xi_0,\omega),\ldots,a_{111}(\theta,\xi_N,\omega)\}, \\
\bar{a}_{217}^T &= \{a_{217}(\theta,\xi_0,\omega),\ldots,a_{217}(\theta,\xi_N,\omega)\}, \\
\bar{a}_{316}^T &= \{a_{316}(\theta,\xi_0,\omega),\ldots,a_{111}(\theta,\xi_N,\omega)\},
\end{aligned}
$$

and the matrix $[c_i a_{ij}]$ is denoted by $\bar{c} \cdot A$, where $A = [a_{ij}]$ $(i,j = 0,\ldots,N)$, and $\bar{c} = \{c_0,\ldots,c_N\}$ the system of differential equations (3.5.11) becomes:

$$
\begin{aligned}
\bar{u}'' &= \Phi_0^{-1}\{(\bar{a}_{12}\cdot\Phi_2 + \bar{a}_{13}\cdot\Phi_1 + \bar{a}_{14}\cdot\Phi_0 + \bar{a}_{111}\cdot\Phi_0)\bar{u} + (\bar{a}_{11}\cdot\Phi_0)\bar{u}' \\
&\quad + (\bar{a}_{17}\cdot X_1 + \bar{a}_{18}\cdot X_0)\bar{v} + (\bar{a}_{15}\cdot X_1 + \bar{a}_{16}\cdot X_0)\bar{v}' + (\bar{a}_{19}\cdot\Psi_1 + \bar{a}_{110}\cdot\Psi_0)\bar{w}\}, \\
\bar{v}'' &= X_0^{-1}\{(\bar{a}_{23}\cdot\Phi_1 + \bar{a}_{24}\cdot\Phi_0)\bar{u} + (\bar{a}_{21}\cdot\Phi_1 + \bar{a}_{22}\cdot\Phi_0)\bar{u}' + \\
&\quad + (\bar{a}_{26}\cdot X_2 + \bar{a}_{27}\cdot X_1 + \bar{a}_{28}\cdot X_0 + \bar{a}_{217}\cdot X_0)\bar{v} + (\bar{a}_{25}\cdot X_0)\bar{v}' \\
&\quad + (\bar{a}_{214}\cdot\Psi_2 + \bar{a}_{215}\cdot\Psi_1 + \bar{a}_{216}\cdot\Psi_0)\bar{w} + (\bar{a}_{2111}\cdot\Psi_2 \\
&\quad + \bar{a}_{212}\cdot\Psi_1 + \bar{a}_{213}\cdot\Psi_0)\bar{w}' + (\bar{a}_{210}\cdot\Psi_0)\bar{w}'' + (\bar{a}_{29}\cdot\Psi_0)\bar{w}'''\}, \\
\bar{w}^{\mathrm{IV}} &= \Psi_0^{-1}\{(\bar{a}_{31}\cdot\Phi_1 + \bar{a}_{32}\cdot\Phi_0)\bar{u} + (\bar{a}_{33}\cdot X_0)\bar{v}' + (\bar{a}_{311}\cdot\Psi_4 + \bar{a}_{312}\cdot\Psi_3 \\
&\quad + \bar{a}_{313}\cdot\Psi_2 + \bar{a}_{314}\cdot\Psi_1 + \bar{a}_{315}\cdot\Psi_0 + \bar{a}_{316}\cdot\Psi_0)\bar{w} + (\bar{a}_{38}\cdot\Psi_2 + \bar{a}_{39}\cdot\Psi_1 \\
&\quad + \bar{a}_{310}\cdot\Psi_0)\bar{w}' + (\bar{a}_{35}\cdot\Psi_2 + \bar{a}_{36}\cdot\Psi_1 + \bar{a}_{37}\cdot\Psi_0)\bar{w}'' + (\bar{a}_{34}\cdot\Psi_0)\bar{w}'''\}
\end{aligned}
\tag{3.5.12}
$$

where:

$$
\begin{aligned}
&u_i^{(k)} = u^{(k)}(\theta,\xi_i); v_i^{(k)} = v^{(k)}(\theta,\xi_i); w_i^{(l)} = w^{(l)}(\theta,\xi_i) \\
&(k = 0,\ldots,1, l = 0,\ldots,3, i = 0,\ldots,N).
\end{aligned}
$$

This system of ordinary differential equations can be represented in normal form:

$$
\frac{d\bar{Y}}{d\theta} = A(\theta,\omega)\bar{Y}, \quad (0 \le \theta \le b), \tag{3.5.13}
$$

where

$$Y^T = \{u_0, \ldots, u_N, u'_0, \ldots, u'_N, v_0, \ldots, v_N, v'_0, \ldots, v'_N, w_0, \ldots, w_N, w'_0, \ldots, w'_N,$$
$$w''_0, \ldots, w'''_N, w'''_0, \ldots, w'''_N\},$$

and $A(\theta, p)$ is a $8(N+1) \times 8(N+1)$ matrix. The boundary conditions (3.5.6)–(3.5.9) for system (3.5.13) can be represented by:

$$B_1\bar{Y}(0) = \bar{0}; \quad B_2\bar{Y}(b) = \bar{0}. \tag{3.5.14}$$

The eigenvalue problem for the system of ordinary differential equations (3.5.13) with the boundary conditions (3.5.14) is solved with the discrete-orthogonalization and incremental search methods (Grigorenko and Mal'tsev [4]). This approach was used to analyze the natural frequency spectrum of isotropic conical panels.

3.5.3 Analysis of the Numerical Results

Consider open isotropic conical shells (panels) defined by eight sets of geometrical parameters collected in Table 3.14, where L is the length of the generatrix, R_1 and R_2 are the radii of the bases; b is the cone angle, and β_k is the taper angle. The thickness of the shells varies according to:

$$h = h_0(1 + \alpha\cos\theta) \quad (-0.5 \le \alpha \le 0.5), \tag{3.5.15}$$

where h_0 is the thickness of a shell of constant thickness and equivalent mass. The shells are made of steel with the following material characteristics: $E = 2.14 \times 10^{11}\,\text{Pa}$, $\nu = 0.2588$, $\rho = 7850\,\text{kg/m}^3$.

The conical panels are subject to the following boundary conditions (3.5.6)–(3.5.9);

Table 3.14 Analytic expressions for the frequencies of a cylindrical panel

L, m	R_1, m	R_2, m	h_0, m	b, rad	β_k, deg	Notation
0.12	0.09	0.09	0.004	$\pi/2$	0	C1
0.12	0.0875	0.0925	0.004	$\pi/2$	2.5	K2.5
0.12	0.085	0.095	0.004	$\pi/2$	5	K5
0.12	0.08	0.1	0.004	$\pi/2$	10	K10
0.12	0.075	0.105	0.004	$\pi/2$	15	K15
0.12	0.07	0.11	0.004	$\pi/2$	20	K20
0.12	0.065	0.115	0.004	$\pi/2$	25	K25
0.12	0.06	0.12	0.004	$\pi/2$	30	K30

G_1 all the edges are clamped (conditions (3.5.6), (3.5.8));
G_2 three edges are clamped (conditions (3.5.6) both (3.5.8) at $\theta = b$) and the fourth edge is hinged (condition (3.5.7) at $\theta = 0$);
G_3 the edges $s =$ const. are clamped (conditions (3.5.6)) and the edges $\theta =$ const are hinged (conditions (3.5.9)).

The frequencies obtained by using the spline-collocation method with different number of collocation points ($N = 10, 12, 14, 16$) differ by no greater than 3 %. Since the frequencies calculated by using $N \geq 12$ collocation points differ slightly, we will discuss the values obtained with $N = 12$.

The reliability of the results is tested by comparing the frequencies of free bending vibrations of a cylindrical shell of constant thickness (C1 in Table 3.14) and conical shells of equivalent mass (K2.5 and K5 in Table 3.14) hinged at all edges. With such a problem formulation, it is possible to derive analytic expressions for the frequencies of a cylindrical panel.

Table 3.15 summarizes the frequencies $\bar{\omega}_i = \omega_i L\sqrt{\rho(1-\nu^2)/E}$ of flexural vibrations of a cylindrical shell and conical shells of equivalent mass with all edges hinged calculated analytically (A) and with the spline-collocation method (B, C, D). Here the frequencies of the cylindrical shell (C1) obtained analytically are in column A; the frequencies for the cylindrical shell (C1) calculated with the spline-collocation method are in column B; the frequencies of the conical shells K2.5 and K5 obtained with the spline-collocation method are in columns C and D, respectively; the relative difference between analytic and numerical results are given in column E. The maximum difference between numerical and analytic results does not exceed 0.4 %, which is indicative of the adequate accuracy of the numerical approach.

The natural frequencies of hinged cylindrical and conical shells with thickness varying as in (3.5.15) were determined as a test problem. The calculated results are collected in Table 3.16, where the frequencies of the cylindrical panel are in column A; the frequencies of the conical panel are in column B; the relative difference between the frequencies of columns A and B are in column P. The difference between the frequencies of the cylindrical and conical panels does not exceed 0.15 %.

Table 3.15 $\bar{\omega}_i = \omega_i L\sqrt{\rho(1-\nu^2)/E}$

i	A	B	E (%)	C	E (%)	D	E (%)
1	0.0777	0.0775	0.36	0.0774	0.46	0.0773	0.61
2	0.1149	0.1147	0.16	0.1147	0.19	0.1146	0.26
3	0.1216	0.1215	0.09	0.1214	0.18	0.1211	0.41
4	0.1572	0.1572	0.01	0.1572	0.04	0.1570	0.16

Table 3.16 $\bar{\omega}_i = \omega_i L\sqrt{\rho(1-\nu^2)/E}$

	α					
	A	B	P (%)	A	B	P (%)
i	−0.2			−0.1		
1	0.0699	0.0698	0.13	0.0724	0.0723	0.11
2	0.0991	0.0991	0.03	0.1061	0.1061	0.03
3	0.1134	0.1133	0.10	0.1134	0.1133	0.11
4	0.1455	0.1454	0.03	0.1502	0.1502	0.03
i	0.1			0.2		
1	0.0776	0.0775	0.09	0.0802	0.0801	0.08
2	0.1135	0.1134	0.10	0.1136	0.1135	0.10
3	0.1198	0.1197	0.03	0.1264	0.1264	0.03
4	0.1586	0.1585	0.02	0.1625	0.1625	0.02

The natural frequencies of conical shells of variable thickness are summarized in Table 3.17 for a cone angle of 10° (K10 in Table 3.14), in Table 3.18 for a cone angle of 20° (K20 in Table 3.14), and in Table 3.19 for a cone angle of 30° (K30 in Table 3.14). Various boundary conditions were examined.

Tables 3.17, 3.18 and 3.19 show how the thickness affects the dynamic characteristics of conical shells and allow analyzing the difference of the natural frequencies of conical shells with variable thickness and constant thickness. This difference increases with increase in the parameter α and in the frequencies.

The frequency is strongly dependent on the parameter α. For example, the frequencies of shells of variable thickness with $|\alpha| = 0.5$ differ from that of a shell of constant thickness and equivalent mass by 12–25 %, depending on the type of boundary conditions. Thus, minor alterations of the thickness may cause considerable changes in the dynamic characteristics of shells comparable (in order of magnitude) with the effect of boundary conditions.

Tables 3.17, 3.18 and 3.19 show how the boundary conditions affect the spectrum of natural frequencies. It can be seen that the more edges are clamped, the higher the frequency will be. In the case of the boundary conditions of type G_3 (edges $s = \text{const}$ are clamped and edges $\theta = \text{const}$ are hinged), the second and third frequencies approach each other as the mid-surface shape tends to that of the conical shell of constant thickness.

For all boundary conditions, the natural frequency of the shell decreases with increasing cone angle. Note that the effect of these changes is an order of magnitude weaker than the effect of the thickness of the shell and the type of boundary conditions. The frequencies of shells with cone angles of 10° and 30° differ

Table 3.17 $\bar{\omega}_i = \omega_i L \sqrt{\rho(1-\nu^2)/E}$

Boundary conditions		α					
G_1	i	−0.5	−0.4	−0.3	−0.2	−0.1	0
	1	0.1017	0.1070	0.1114	0.1155	0.1194	0.1233
	2	0.1052	0.1126	0.1201	0.1273	0.1340	0.1402
	3	0.1568	0.1671	0.1756	0.1831	0.1903	0.1973
	4	0.1589	0.1692	0.1787	0.1875	0.1951	0.2015
	i	0.1	0.2	0.3	0.4	0.5	0
	1	0.1271	0.1309	0.1347	0.1385	0.1422	–
	2	0.1460	0.1513	0.1561	0.1604	0.1642	–
	3	0.2042	0.2102	0.2149	0.2196	0.2242	–
	4	0.2066	0.2119	0.2181	0.2236	0.2284	–
G_2	i	−0.5	−0.4	−0.3	−0.2	−0.1	0
	1	0.0880	0.0904	0.0933	0.0957	0.0982	0.1007
	2	0.0982	0.1057	0.1124	0.1191	0.1251	0.1307
	3	0.1360	0.1456	0.1512	0.1548	0.1576	0.1608
	4	0.1526	0.1604	0.1661	0.1710	0.1756	0.1802
	i	0.1	0.2	0.3	0.4	0.5	0
	1	0.1032	0.1053	0.1078	0.1102	0.1127	–
	2	0.1353	0.1392	0.1424	0.1449	0.1473	–
	3	0.1647	0.1689	0.1742	0.1798	0.1859	–
	4	0.1844	0.1887	0.1926	0.1964	0.2000	–
G_3	i	−0.5	−0.4	−0.3	−0.2	−0.1	0
	1	0.0770	0.0806	0.0837	0.0869	0.0901	0.0929
	2	0.0883	0.0954	0.1025	0.1096	0.1167	0.1226
	3	0.1194	0.1201	0.1208	0.1212	0.1219	0.1238
	4	0.1283	0.1417	0.1533	0.1611	0.1682	0.1742
	i	0.1	0.2	0.3	0.4	0.5	0
	1	0.0957	0.0989	0.1018	0.1046	0.1074	–
	2	0.1230	0.1237	0.1244	0.1251	0.1261	–
	3	0.1308	0.1377	0.1445	0.1514	0.1582	–
	4	0.1798	0.1851	0.1901	0.1947	0.1989	–

by 5–7 %. The greater the cone angle, the sharper the change of the natural frequencies.

Figure 3.14 shows mode shapes corresponding to four natural frequencies of shells with one edge hinged and three edges clamped. As can be seen, the first natural frequency corresponds to two circumferential half-waves and one

Table 3.18 $\bar{\omega}_i = \omega_i L\sqrt{\rho(1-\nu^2)/E}$

Boundary conditions		α					
G_1	*i*	−0.5	−0.4	−0.3	−0.2	−0.1	0
	1	0.0999	0.1049	0.1092	0.1134	0.1174	0.1213
	2	0.1039	0.1114	0.1188	0.1258	0.1323	0.1383
	3	0.1538	0.1635	0.1719	0.1797	0.1871	0.1934
	4	0.1553	0.1670	0.1769	0.1847	0.1895	0.1945
	i	0.1	0.2	0.3	0.4	0.5	0
	1	0.1252	0.1290	0.1329	0.1368	0.1406	–
	2	0.1437	0.1486	0.1530	0.1569	0.1604	–
	3	0.1976	0.2020	0.2066	0.2116	0.2167	–
	4	0.2016	0.2087	0.2150	0.2200	0.2247	–
G_2	*i*	−0.5	−0.4	−0.3	−0.2	−0.1	0
	1	0.0862	0.0890	0.0915	0.0940	0.0965	0.0989
	2	0.0972	0.1046	0.1113	0.1177	0.1237	0.1286
	3	0.1328	0.1417	0.1466	0.1502	0.1530	0.1565
	4	0.1512	0.1579	0.1632	0.1682	0.1728	0.1774
	i	0.1	0.2	0.3	0.4	0.5	0
	1	0.1014	0.1039	0.1064	0.1088	0.1113	–
	2	0.1328	0.1364	0.1392	0.1420	0.1442	–
	3	0.1608	0.1657	0.1710	0.1770	0.1834	–
	4	0.1820	0.1859	0.1901	0.1936	0.1975	–
G_3	*i*	−0.5	−0.4	−0.3	−0.2	−0.1	0
	1	0.0753	0.0791	0.0823	0.0855	0.0883	0.0915
	2	0.0876	0.0947	0.1018	0.1092	0.1163	0.1187
	3	0.1152	0.1162	0.1170	0.1173	0.1180	0.1233
	4	0.1261	0.1392	0.1505	0.1583	0.1650	0.1714
	i	0.1	0.2	0.3	0.4	0.5	0
	1	0.0947	0.0975	0.1007	0.1035	0.1067	–
	2	0.1194	0.1201	0.1208	0.1215	0.1226	–
	3	0.1300	0.1371	0.1438	0.1505	0.1572	–
	4	0.1774	0.1827	0.1876	0.1922	0.1964	–

longitudinal half-wave; the second frequency to three circumferential half-waves and one longitudinal half-wave; the third frequency to one longitudinal half-wave and four circumferential half-waves; and the fourth frequency corresponds to two circumferential half-waves and two meridional half-waves.

Table 3.19 $\bar{\omega}_i = \omega_i L\sqrt{\rho(1-\nu^2)/E}$

Boundary conditions		α					
G_1	i	−0.5	−0.4	−0.3	−0.2	−0.1	0
	1	0.0960	0.1009	0.1053	0.1095	0.1136	0.1176
	2	0.1014	0.1089	0.1160	0.1226	0.1286	0.1340
	3	0.1472	0.1568	0.1646	0.1703	0.1748	0.1793
	4	0.1516	0.1631	0.1686	0.1749	0.1823	0.1898
	i	0.1	0.2	0.3	0.4	0.5	0
	1	0.1217	0.1257	0.1296	0.1336	0.1376	–
	2	0.1387	0.1429	0.1466	0.1498	0.1526	–
	3	0.1842	0.1893	0.1948	0.2005	0.2063	–
	4	0.1972	0.2031	0.2079	0.2125	0.2173	–
G_2	i	−0.5	−0.4	−0.3	−0.2	−0.1	0
	1	0.0832	0.0859	0.0885	0.0912	0.0938	0.0964
	2	0.0952	0.1025	0.1092	0.1153	0.1206	0.1249
	3	0.1276	0.1344	0.1385	0.1418	0.1454	0.1497
	4	0.1466	0.1526	0.1578	0.1630	0.1680	0.1726
	i	0.1	0.2	0.3	0.4	0.5	0
	1	0.0990	0.1015	0.1039	0.1063	0.1086	–
	2	0.1284	0.1312	0.1337	0.1360	0.1383	–
	3	0.1548	0.1605	0.1666	0.1729	0.1794	–
	4	0.1770	0.1811	0.1850	0.1888	0.1925	–
G_3	i	−0.5	−0.4	−0.3	−0.2	−0.1	0
	1	0.0724	0.0760	0.0795	0.0827	0.0858	0.0890
	2	0.0866	0.0936	0.1011	0.1081	0.1109	0.1116
	3	0.1078	0.1088	0.1095	0.1102	0.1152	0.1223
	4	0.1230	0.1357	0.1456	0.1530	0.1601	0.1668
	i	0.1	0.2	0.3	0.4	0.5	0
	1	0.0922	0.0954	0.0986	0.1018	0.1046	–
	2	0.1124	0.1131	0.1141	0.1152	0.1162	–
	3	0.1293	0.1360	0.1427	0.1495	0.1562	–
	4	0.1731	0.1788	0.1834	0.1876	0.1915	–

These results are indicative of the high efficiency of the spline-collocation method when applied to analyze the natural frequency spectrum of thin conical shells (panels) with arbitrarily varying thickness.

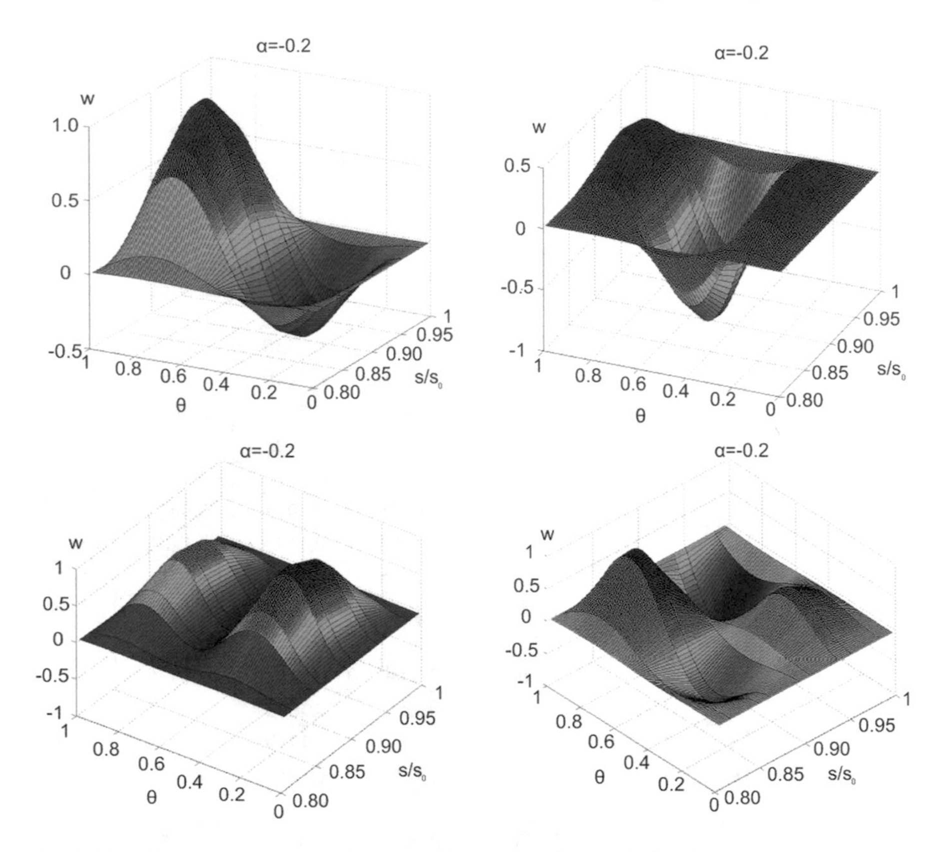

Fig. 3.14 Effect of boundary conditions on the modes of natural vibrations

References

1. Ding K, Tang L (1999) The solution of weak formulation for axisymmetric problem of orthotropic cantilever cylindrical shell. Appl Math Mech 20(6):615–621
2. Donnell LG (1976) Beams, plates and shells. McGraw-Hill, New York
3. Flügge W (1967) Stresses in shells. Springer, Berlin
4. Grigorenko AY, Mal'tsev SA (2009) Natural vibrations of thin conical panels of variable thickness. Int Appl. Mech. 45(11):1221–1231
5. Grigorenko AY, Tregubenko TV (2000) Numerical and experimental analysis of natural vibration of rectangular plates with variable thickness. Int Appl Mech 36(2):268–270
6. Grigorenko YM (1973) Izotropnyye i anizotropnyye sloistyye obolochki vrashcheniya peremennoy zhestkosti (Isotropic and anisotropic laminated shells of revolution with variable stiffness). Naukova Dumka, Kiev
7. Grigorenko YM, Grigorenko AY, Zakhariichenko LI (2005) Stress-Strain analysis of orthotropic closed and open noncircular cylindrical shells. Int Appl Mech 41(7):778–785
8. Grigorenko YM, Grigorenko AY, Zakhariichenko LI (2006) Stress-strain solutions for circumferentially corrugated elliptic cylindrical shells. Int Appl Mech 42(9):1021–1028

9. Grigorenko YM, Vasilenko AT (1997) Solution of problems and analysis of the stress-strain state of nonuniform anisotropic shells (survey). Int Appl Mech 33(11):851–880
10. Grigorenko YM, Zakhariichenko LI (1998) Solution of the problem of the stress state of noncircular cylindrical shells of variable thickness. Int Appl Mech 34(12):1196–1203
11. Johns IA (1999) Analysis of laminated anisotropic plates and shells using symbolic computation. Int J Mech Sci 41(4–5):397–417
12. Korn GA, Korn TM (1968) Mathematical handbook for scientists and engineers. Definitions, theorems, and formulas for reference and review. McGraw-Hill, New York
13. Li L, Babuska I, Chen J (1997) The boundary buyer for p–model plate problems. Pt 1. Asymptotic analysis. Acta Mech 122(1–4):181–201
14. Librescu L, Hause T (2000) Recent developments in the modeling and behavior of advanced sandwich constructions (survey). Compos Struct 48:1–17
15. Librescu L, Schmidt R (1991) Substantiation of a shear deformable theory of anisotropic composite laminated shells accounting for the interminated continuity conditions. Int J Eng Sci 29(6):669–683
16. Noor AK, Burton WS (1992) Computational models for hightemperature multilayered composite plates and shells. Appl Mech Rev 45(10):419–446
17. Ramm E (1977) A plate/shell element for large deflections and rotations. In: Bathe A et al (eds) Formulations and computational algorithms in finite element analysis. MIT Press, Cambridge
18. Soldatos KP (1999) Mechanics of cylindrical shells with non-circular cross-section. A survey. Appl Mech Rev 52(8):237–274
19. Sun BH, Zhang W, Yeh KY, Rimrott FDJ (1996) Exact displacement solution of arbitrary degree paraboidal shallow shell of revolution made of linear elastic materials. Int J Solids Struct 33(16):2299–2308
20. Vlasov VZ (1964) General theory of shells and Its application to engineering. NASA Technical Translation, NASA-TT-F-99, Washington
21. Zang R (1999) A novel solution of toroidal shells under axisymmetric loading. Appl Math Mech 20(5):519–526

Conclusions

Methods of analysis of the mechanical behavior of anisotropic inhomogeneous shell structures was proposed. As a first mechanical model, the classic Kirchhoff-Love shell theory, the Timoshenko-Mindlin refined theory, and three-dimensional elasticity theory were used. Approaches to the solution of linear problems of mechanics (statics and dynamics) of shells on the basis of discrete-continual methods in classical, refined, and tree dimensional formulations for anisotropic inhomogeneous shells with variable geometrical and mechanical parameters were presented. The developed approaches were realized in form of software-based algorithms. Advantages of the method include: reduction of the partial differential equations to one dimensional problems (spline collocation method, the discrete Fourier series method) and exact treatment of boundary conditions, and practically, the exact solution of boundary value problems and eigenvalue problems for systems of ordinary differential equations with variable coefficients (discrete orthogonalization method). The results of investigations of the stress state and dynamic characteristics of shells of various shape and structure were analyzed depending on the variation of basic parameters. Calculations of specific structural members were carried out.

The following problems based on the classical model were solved:

- the stress-strain state of multilayer shallow rectangular in-plane shells that are symmetrical about the mid-surface of the structure and composed of an odd number of orthotropic layers of variable thickness;
- the stress-strain state in noncircular isotropic and anisotropic thin cylindrical shells with a thickness varying along a generatrix and directrix under arbitrary surface load; an elliptic cross-section of noncircular cylindrical shells and corrugated cylindrical shells was considered;
- the stress state of a high-pressure balloon made of a glass-reinforced plastic;
- the free vibration of orthotropic rectangular plates of variable thickness under different boundary conditions;
- the free vibrations of truncated conical shells with a circumferentially inhomogeneous cross-section;

  the effect of variation in thickness, mechanical parameters, and boundary conditions and the type of the loads on the behavior of displacement, stresses,

A.Ya. Grigorenko et al., *Recent Developments in Anisotropic Heterogeneous Shell Theory*, SpringerBriefs in Continuum Mechanics,
DOI 10.1007/978-981-10-0353-0

natural frequencies, and vibration modes of the anisotropic inhomogeneous shells structures was analyzed.

- the reliability of the obtained results is proved by means of an inductive technique, by comparison with test examples for isotropic materials and some types of boundary conditions, and also by comparison with experimental data. The possibility of application of the developed approach to the solution of a new class of problems related to the mechanical behavior of a wide class of shell constructions made of modern anisotropic layered and continuously-inhomogeneous materials is shown.

Printed in the United States
By Bookmasters